How Do I AI? For Real Estate Agents and Brokers

The Practical Guide to Becoming an AI-Amplified Agent: Reclaim Your Time, Multiply Your Impact, and Build a Sustainable Business

Jim Washok

Published by WOW NOW Media

ISBN (Paperback): 979-8-9943434-0-1
ISBN (eBook): 979-8-9943434-1-8

First Edition: January 2026

WOW NOW Media
Richmond, VA
howdoiai.pro
jimwashok.com

Dedication

I dedicate this book first to you, the reader — the real estate agents and brokers who serve as the backbone of our communities. I know the weight you carry. You are often singularly responsible for running an entire business while remaining on call... days, nights, and weekends... for clients navigating the largest financial transactions of their lives.

You don't just sell shelter; you help people find the homes where society takes root and memories are made. From the moment I began writing, honoring that grind was my priority. I worked hard to filter out the noise and focus only on what you need to build a sustainable, thriving business. I hope I did you right.

To my amazing wife Donna, who has chosen to love and support me every day for over 31 years. To my three incredible 'kids' — Kailey, Brianna, and Jordan — though they are all grown up are as fun as ever to hang out with along with my son-in-law Joey. And to my grandson, who fills my heart with joy every time I see him.

To my loving mom, Tia, and dad, Jim, who always had my back and took such great care of me. Much love as well to my stepmother, stepfather, stepbrothers, and stepsisters.

To my 'extended family' at Gather, Heights Church, WEAG Church, and Hope Church: thank you for years of encouragement, great laughs, and late nights tackling impossible missions.

To my friends Erin, John, Trevor, Ben, and Brian — the OGs of our AI workshop. You inspired this book and helped me amplify my own work. To my many friends in the real estate industry who provided the insights I sought to address in these pages.

Most importantly, I dedicate this work to my Lord and Savior Jesus Christ, who has blessed my life in countless ways and continues to prove His love for me and my family over and over.

Foreword: A Confession, an Invitation, and a Few Key Notes

I need to tell you something before you turn another page: this book was built with AI.

It wasn't written by AI, but was heavily, unapologetically, and intentionally built with it. To bring this resource to life, I leveraged AI to:

- **Strategize:** Conduct research, assess storyline ideas, and validate story arcs.

- **Draft:** Consider potential outlines, craft rough drafts of every section, and variations to consider of what I wanted to say.

- **Refine:** Evaluate completeness and tone, look up sources, and reword my edits.

- **Edit:** Brutally critique what worked and discard what fell short.

- **Test:** Iterate on title concepts and role-play potential reader reactions, interpretations, and applicable take-aways.

- **Build:** Code a custom utility app from scratch based on my specs to automate the bulk of formatting of this book for publishing.

Why would I admit this? Because the whole point of what you're about to read is that AI can amplify human potential. I would have been doing you a disservice to not demonstrate exactly that in creating this resource for you. When the reality is I use AI to amplify almost everything I do.

Here's what I discovered while simultaneously working on this book and another volume completely unrelated to real estate or AI: the process was both revealing and enabling in ways I didn't fully anticipate. I'm not a great wordsmith. Verbosity is all too natural for me. I sometimes read my own sentences and end up bewildered trying to decipher what I was trying to say.

The helpfulness of my AI tools to take my swirling, loosely-threaded-together thoughts and reorient them into something more eloquent, readable, and understandable felt like I was Tony Stark stepping into an

Iron Man super-exoskeleton. Even this very paragraph you're reading now was likely quite different when I first wrote it, and almost certainly with many more words than necessary to get the point across.

Could I have done all this without AI? Sure. Research, storyline testing, and source citation were all possible using search engines before generative AI emerged. But the enormous number of hours it would have taken to conduct searches and dig through every promising link for this book, plus three others at various levels of completion, all while servicing my business clients, would mean this resource wouldn't be available to you right now when you need it. It would be many months, possibly even years, off in the distance. AI is changing way too fast to keep you waiting that long.

If you are uncomfortable with my heavy collaboration with AI in this process, I must ask: which matters more? The process by which every word was chosen for inclusion in this book? Or the newfound process by which you can achieve greater success by implementing what I've consolidated here while avoiding burn-out? I contend that what really matters is *not what I did* but *what you will do* with what I did.

I can assure you the text you're about to read is valid, unique, and deeply personal while being about you, not me. Everything AI helped with was heavily directed by, influenced by, and refined by me. This process was not as simple as giving ChatGPT a prompt: *"Create a 250+ page book that teaches real estate agents and brokers how to use AI to save time performing their jobs."* If it were that simple, I wouldn't have needed to spend so many hours to craft the insights within. I could have just posted the magical prompt to social media for you to execute on your own.

The reality is any and every AI platform currently available would choke on that prompt. The result would be incredibly disappointing. It would have been full of how-tos but lacking in story. It would have presented research, some of it possibly incorrect in whole or in part, while being void of context. It probably would have been overly technical and address current specifics of software which would have left you lacking in instruction that is applicable beyond the next year or two. Rather, with this book you will benefit from my 30+ years of designing, building, and supporting hundreds of technical implementations of all sizes and types.

If I… and you… can benefit from *AI-amplified authorship,* then you… and those you help… can benefit from you being an *AI-amplified agent*!

Before you dive in, a few important notes:

Agents is a term that applies quite commonly to both real estate and AI. In all cases, the use of 'agent' or 'agents' alone will refer to the role of a real estate agent and the persons in such roles. When I refer to agents in the context of AI assistants and automations acting on behalf of human interests, I will preface the term with 'AI' as in 'AI agent' and 'AI agents'.

You will find that I make many mentions of AI weaknesses and concerns over inaccurate and incomplete results as you utilize AI in your work, especially as it relates to the regulated areas of real estate. I'm sorry if my warnings seem overly frequent and redundant. That's purposeful. AI is awesome, but it is not infallible. I don't want you to pay a hefty price for a misuse that I didn't responsibly warn you about.

You will not find screenshots of tools and highly detailed step-by-step instructions in this book. For one, I didn't want it to be a software user manual. That's boring. Additionally, the software I introduce is changing too quickly, and I want this book to be a resource for you for years to come. While I do detail how to use different types of AI tools, you'll find that the process I outline is a few levels above specific steps. Rather, I aimed to guide you in what to look for and achieve the result you're aiming for with a particular *type of* tool versus a *specific* tool.

Two years from now, some of the tools I list might not be available anymore. There will be others in their place, and the workflow outcome will still be valid. If you're picking this book up years from publication, I don't want it to immediately feel dated. While Tool ABC that I identify may not be available anymore, there is surely a similar Tool XYZ which you can use instead.

Not only might the available tools probably differ from the time of publication till the time you read this book, but tool functionality, packages, pricing, and URLs might also be different. I encourage you to always verify the latest about a tool before signing up and committing to it.

Now, let's talk about you and the AI-amplified future waiting for you.

Contents

Contents of Appendices

Introduction: Your Path to Becoming an AI-Amplified Agent

Picture your typical week as a real estate agent.

Monday morning starts with 47 unread emails, 13 text messages, and three voicemails from the weekend. A buyer who toured five properties on Saturday is now asking detailed questions about school districts, HOA fees, and resale potential for each one. A listing client wants drone footage, a 3D tour, professional staging photos, and a social media campaign ready by Thursday. Your phone buzzes with a Zillow lead notification. Then another. Then three more.

By Tuesday afternoon, you're writing property descriptions, researching comparable sales, scheduling showings, following up with mortgage lenders, updating your CRM, and trying to figure out why your Instagram engagement dropped 40% last month. Wednesday brings a contract with 17 pages that need review before tomorrow's deadline. Thursday you're filming a neighborhood tour video for TikTok because "that's where the buyers are now." Friday you're answering the same first-time buyer questions you've answered 200 times before.

Saturday and Sunday? Showings, open houses, and the guilty feeling that you should be spending this time with your family, or maybe finally taking that vacation you've been postponing for two years.

Sound familiar?

This was the reality I kept hearing about two years ago when agents I knew started asking about how I use AI to amplify my efforts. As someone who has spent over 30 years deploying technical solutions for professionals and service-based companies, I've always been fascinated by technology and drawn to its potential to empower people in ways previously only imagined. When multiple real estate agents began reaching out for guidance on AI tips, tools, and techniques, I realized something significant was happening.

The pressure on agents had reached unsustainable levels. Not just the timeless challenge of selling and buying in competitive markets, but something more complex and more exhausting. Economic headwinds had created a perfect storm: mortgage rates climbing to levels not seen in decades, inventory constraints making every listing more competitive, and housing affordability stretching beyond reach for many buyers.

Meanwhile, the digital marketing landscape had fractured into dozens of platforms, each demanding fresh content, authentic engagement, and platform-specific strategies. What worked in 2020 already felt obsolete. The tools and methods from 2015? Those were artifacts from a different era entirely.

I started this project because talented, dedicated agents were drowning in demands that had become humanly impossible to meet through sheer effort alone. And I knew that artificial intelligence, properly understood and thoughtfully applied, could change everything.

Who This Book Is For (And Who It Isn't)

This book is for you if:

- You feel overwhelmed by administrative work that steals time from clients and family
- You know AI is important but don't know where to start
- You've tried AI tools but couldn't make them work consistently
- You want to serve more clients without sacrificing quality or burning out
- You're worried about falling behind tech-savvy competitors
- You value practical solutions over theoretical possibilities

This book is NOT for you if:

- You're looking for get-rich-quick schemes or "10x your income overnight" promises
- You want to automate away client relationships and human connection
- You expect AI to replace your professional judgment and expertise

- You're seeking detailed programming tutorials or technical deep-dives
- You're not willing to invest time learning new approaches

I'm not here to turn you into a technologist. I'm here to show you how to use technology the way successful agents already do: as a teammate that handles grunt work so you can do what you're great at and prefer to do.

What Makes This Book Different

I'm not a real estate agent. I've never sold a property professionally. I've never negotiated a contract to close a deal for housing on behalf of a client.

It proved beneficial to be on the outside of the problem looking in. I could see clearly what could work based on my experience and prior results with clients in similar predicaments.

I've spent three decades implementing technology solutions for professionals and process-heavy organizations across industries, including construction, insurance, financial services, and yes, real estate. I've seen what works and what fails. I've watched successful implementations and spectacular crashes.

Here's the pattern I've observed: **The best technology implementations don't replace nor get in the way of human expertise. They amplify it.**

The doctors who thrive with electronic health records weren't the most technical doctors. They were the ones who used technology to spend more face time with patients, not less. The attorneys who excel with AI-powered legal research weren't the ones who automated away their judgment. They were the ones who used efficiency gains to provide deeper strategic counsel.

Real estate follows the same pattern. The agents winning with AI aren't replacing human connection with automation. They're using automation to create space for more meaningful human connection.

That's the philosophy guiding every chapter of this book: **AI as teammate, not replacement. Efficiency as means, not end. Technology enabling humanity, not diminishing it.**

You'll find:

- **No jargon without explanation.** When technical terms are necessary, I define them in plain English immediately.
- **No theoretical concepts without practical application.** Every chapter includes specific implementations you can use today.
- **No tool recommendations without budget context.** I prioritize affordable, high-value solutions and tell you exactly what premium tools cost as of the time of writing.
- **No efficiency tactics without ethics.** Fair Housing compliance, data privacy, and professional responsibility are woven throughout, not afterthoughts.
- **No promises without measurement.** Chapter 7 teaches you how to calculate actual ROI and prove value.

What You'll Learn

This book provides a complete roadmap for integrating AI into your real estate practice, regardless of your current technical skill level.

Chapters 1-2 lay the foundation of what AI is, how it thinks, and the core concepts you need to understand. You'll set up your first tools and complete your first successful AI-assisted tasks.

Chapters 3-4 help you identify where AI delivers the highest value in your specific business and how to select tools that match your budget and needs. You'll avoid expensive mistakes and build a cost-effective toolkit.

Chapter 5 provides detailed, copy-paste-ready workflows for every major real estate task: property marketing, client communications, research and analysis, transaction management, and business development. This is the tactical heart of the book.

Chapter 6 elevates your thinking from individual tasks to connected systems, such as automation workflows that create seamless client experiences while multiplying your capacity.

Chapter 7 teaches measurement and ROI calculation so you can prove (to yourself and others) that your AI investment is delivering tangible results. You'll track what matters and ignore vanity metrics.

Chapter 8 provides troubleshooting frameworks for when things inevitably don't work perfectly. You'll learn to diagnose root causes and implement targeted solutions rather than abandoning AI at the first obstacle.

Chapter 9 prepares you for AI's continued rapid evolution over the years to come. You'll develop the adaptability mindset that ensures you thrive regardless of which specific tools win or lose.

Chapter 10 creates your personalized action plan. Not some universal prescription, but a realistic roadmap that acknowledges your actual constraints and priorities.

Three appendices provide ready-to-use prompt templates, current tool comparisons with pricing, and a comprehensive glossary of essential terms.

The AI-Amplified Agent

Throughout this book, you'll encounter the concept of the "AI-Amplified Agent" (or interchangeably the "Amplified Agent"). The AI-Amplified Agent is a professional agent or broker who uses AI not just to work faster (though that happens), but to work differently at a fundamental level.

Today, you may spend 60% of their time on administrative tasks, 25% on client interaction, 10% on business development, and 5% on personal life and growth.

As an AI-Amplified Agent, you'll flip that to 20% administrative (AI-assisted and efficient), 40% high-value client interaction, 20% strategic business growth, and 20% life margin for family, health, and the pursuits that matter to you more than your production numbers.

Same 40-hour workweek. Radically different allocation of your energy.

The difference isn't working harder. It's not having better natural talent or more resources. It's systematically deploying AI to handle what it does well (consistent execution of routine tasks) while focusing your irreplaceable human capabilities where they matter more (judgment, relationship-building, creative problem-solving, emotional intelligence, and local expertise).

This book shows you how to become that amplified agent.

Not six years from now when you've "figured it all out." Not after you become some kind of technology expert. Starting in the next 30 days, with the constraints and resources you have right now.

A Note on Change

I know this feels like a lot. Adding AI to your practice probably sounds overwhelming on top of everything else demanding your attention.

But here's what I've learned from watching hundreds of agents navigate this transition: **The overwhelm comes from trying to do everything at once.**

You don't need to implement everything in this book. You don't need to master every tool or build every workflow. Just start with one thing that genuinely solves a real problem you have right now.

For some agents, that's automated listing descriptions that save 90 minutes weekly. For others, it's email follow-up systems that ensure no lead is forgotten. For others, it's transaction coordination that eliminates the 2 AM panic about missed deadlines.

One meaningful change, implemented well, beats ten perfect plans that never happen.

Chapter 10 will help you identify your one next step. For now, just know that this is achievable. You don't need to be a technical whiz. You don't need unlimited time or budget. You just need to be willing to try something new and give it a fair chance to prove value.

You may feel overwhelmed, uncertain, maybe a little skeptical. That's all understandable.

But I encourage you to just take a step. Then another. Then another.

Six months down the road and you're sure to be serving more clients, earning more income, working fewer hours, and enjoying their careers again instead of just surviving them.

You can do this. This book will show you how.

Let's begin.

Chapter 1: AI as Your Real Estate Sidekick

A Tuesday That Changed Everything

Sarah Mitchell closed her laptop at 9:47 PM and stared at the stack of unsigned listing agreements on her desk. Three potential sellers waiting for customized market analyses. Seventeen unread emails from active buyers. A showing feedback form she'd promised to send yesterday. And tomorrow, she'd miss her son's basketball game. Again.

She was closing 18 deals a year (respectable for a solo agent) but the cost was mounting. Missed family dinners. Delayed vacations. A growing sense that she was running faster just to stay in place.

A year later, Sarah closed 22 deals. But she worked 7 fewer hours per week. She made every basketball game that season. She took a two-week vacation without her phone buzzing constantly. Her clients rated her responsiveness higher than ever before.

What changed wasn't her work ethic or her market expertise. What changed was her relationship with artificial intelligence.

The AI Revolution in Real Estate

The real estate industry stands at a defining moment. Artificial intelligence has moved beyond experimental novelty to become a fundamental capability that separates thriving practices from struggling ones. This isn't happening in some distant future. It's happening right now in markets across the country.

The data tells a compelling story. Real estate companies strategically deploying AI have realized net operating income gains of 10% or more through superior operating models, enhanced customer experiences, and smarter decision-making. Industry analysts project that AI could drive $34 billion in efficiency gains for the real estate sector by 2030.

But here's what those numbers don't capture: AI isn't just about efficiency metrics or profit margins. It's about margin for life. It's about serving more clients without sacrificing service quality. It's about building a practice that generates wealth and creates space for what matters most: family, friends, fitness, faith, and the pursuits that make you more than your production numbers.

The reality facing real estate professionals right now: AI won't replace agents, but agents who effectively use AI will almost certainly outperform those who don't. The competitive gap is already visible. The question isn't whether AI will reshape real estate practices (it is already making an impact). The question is whether you'll use it to become more of who you are at your best.

The AI-Amplified Agent Framework

Think about how you currently spend your time. If you're like most agents, your week breaks down something like this:

Traditional Agent Time Allocation:

- 60% Administrative work and grunt tasks (descriptions, emails, research, paperwork)
- 25% Client interaction (showings, consultations, negotiations)
- 10% Business development (prospecting, networking, marketing)
- 5% Personal development and life margin

Now imagine a different allocation:

The Amplified Agent Time Allocation:

- 20% Administrative work (AI-assisted, streamlined, faster)
- 40% High-value client interaction (relationship building, strategic guidance)
- 20% Strategic business growth (the work that scales your practice)

- 20% Life margin (family time, health, community, rest)

This is working fundamentally differently. Work that is focused much more on desired outcomes than on process performance.

The Amplified Agent isn't defined by:

- The number of AI tools in their tech stack
- Their technical sophistication
- Their ability to discuss machine learning algorithms

The Amplified Agent is defined by:

- Using AI to reclaim time for high-value human activities
- Serving more clients without diminishing service quality
- Building systems that create both professional success and personal margin
- Focusing their irreplaceable human capabilities where they matter most

This book provides the roadmap for that transformation.

How This Guide Works

This isn't a technical manual. It's a strategic playbook designed for busy real estate professionals who need practical solutions, not theoretical possibilities.

I've structured this guide to meet you exactly where you are:

Complete Beginners (Never used AI tools before):

Start with Chapter 2: AI Foundations for Real Estate Professionals. I'll walk you through setup, terminology, and your first implementations. Look for **[BEGINNER]** tags throughout the book.

Some AI Experience (Experimented with ChatGPT or similar tools):

You may prefer to skip to Chapter 3: Daily Workflow Integration, Pay particular attention to content marked with **[INTERMEDIATE]** tags. I'll help you systematize what you've been doing ad hoc.

Experienced AI Users (Already using AI but want real estate-specific optimization):

You are welcome to jump ahead to Chapter 5: Task-Specific Applications for detailed workflows across every aspect of your practice. Or perhaps skim the preceding chapters for any sections you feel you could use a bit of a refresher or extra insights on before kicking off your journey with chapter 5.

What This Guide Is (and Isn't)

This guide IS:

- A practical roadmap for incorporating AI into your existing real estate business
- A collection of proven strategies specific to real estate workflows
- A balanced approach valuing your human expertise while leveraging technological efficiency
- An ethical framework for using AI responsibly in a regulated industry
- A vision for building a practice that creates both wealth and life margin

This guide IS NOT:

- A promise that AI will "10x your business overnight" without strategic effort
- A technical deep dive into programming, coding, or development
- A comprehensive exploration of every AI tool on the market
- A suggestion to blindly automate every aspect of your business
- A replacement for your professional judgment, local expertise, and client relationships

Real estate remains a profession built on trust, relationships, and local knowledge (elements no AI can replicate). My goal is to amplify these distinctly human advantages by removing the administrative burden that steals time from your highest-value activities.

The Prime Directive: Reclaiming Your Life Margin

Before I discuss tools, workflows, or implementation strategies, let's establish the fundamental purpose of AI in your practice: **reclaiming margin for life**.

Margin isn't just about working fewer hours (though that's certainly valuable). Margin is about:

- Having the mental clarity to be fully present with clients during critical moments
- Building relationships that generate referrals rather than constantly chasing cold leads
- Investing in your health without feeling guilty about time away from work
- Being present for family moments that matter: recitals, games, dinners, conversations
- Pursuing interests and relationships that make you more than your job title
- Building a practice you can sustain for decades, not just survive for years

Real estate agents spend significant time on tasks that don't directly generate leads or close deals: paperwork, email management, follow-ups, marketing logistics, and basic research. Some agents have shared that as many as 30 of the 40 typical hours required to close a deal are spent on administrative tasks.

Imagine reclaiming even half that administrative time. That's 260+ hours annually (equivalent to six full 40-hour workweeks).

An extra month and a half every year.

What would you do with that time?

Would you:

- Serve more clients without increasing stress?
- Finally launch that content marketing strategy you've been planning?
- Take actual vacations where you're not constantly checking email?
- Invest in deeper relationships with your sphere of influence?

- Coach your kid's team or volunteer in your community?
- Build the investment portfolio you keep postponing?

The answer to that question should guide every AI decision you make. AI that doesn't create margin (either for business growth or personal life) isn't worth implementing.

The Irreplaceable You: What AI Can't Touch (And Never Will)

As I explore AI's capabilities throughout this book, never lose sight of a fundamental truth: the most valuable aspects of real estate practice remain well beyond AI's reach.

The Empathy Advantage

Picture this: A buyer calls you in tears. Her financing fell through two days before closing. She's already given notice at her apartment, and they have a new tenant moving in. She's panicking. Her voice is shaking.

No chatbot can talk her off that ledge of worry that everything is falling apart. No algorithm can convey the calm confidence that comes from experience. No AI can provide the human reassurance that says, "I've navigated this exact situation before, and here's exactly what we're going to do."

That moment of human connection during crisis is where your true value crystallizes. AI can draft the follow-up emails and coordinate the backup offer. But only you can provide the steady presence your client desperately needs.

The Intuition Factor

You walk into a listing presentation. The homeowners say they want top dollar and a quick sale. But something in their body language tells you a different story. They're nervous. They're uncertain. There's tension between the spouses about timing.

Your experience tells you to slow down. To ask different questions. To address the unspoken concerns before talking about marketing strategy.

That human intuition developed through hundreds of client interactions cannot be programmed. AI can help you prepare the most sophisticated market analysis ever created. But it can't tell you when to set the data aside and simply listen.

The Trust Multiplier

Why do clients hire you instead of a discount broker or a tech platform? Because real estate transactions are high-stakes, emotionally charged, and complex. People need someone they trust to guide them through the biggest financial decision of their lives.

That trust isn't built through efficiency or data analysis. It's built through:

- Showing up consistently over months or years of relationship building
- Demonstrating integrity in small decisions that reveal character
- Providing value without expectation of immediate return
- Being present and attentive during moments that matter
- Following through on commitments, especially when it's inconvenient

AI can help you maintain consistent communication and follow-up. But it cannot build trust. Only you can do that.

The Community Connector

You know things about your market that no dataset captures. You know which neighborhoods have the strongest sense of community. You know which school principal has transformed the local elementary school. You know that the house on Maple Street looks perfect but has drainage issues every spring. You know which streets get plowed first in winter. You know where the new grocery store is going before it's announced.

This hyperlocal, deeply contextual knowledge is your competitive moat. AI can process MLS data and analyze trends. But it can't tell your buyer that their kids will grow up playing with the neighbor's children because you sold that house to a young family three years ago.

By 2030, the most valuable agents won't be those who know the most AI tools. They'll be the ones who've used AI to become the most human: the most present, the most empathetic, the most trusted. While competitors drown in data, you'll be the calm, confident guide clients desperately need.

A Critical Note on AI Capabilities and Limitations

AI tools are powerful assistants, but they are not infallible. They can generate outdated information, make factual errors, produce content requiring significant refinement, or confidently state things that are false (a phenomenon called "hallucination").

Every AI-generated output in your business must be reviewed by you before client delivery.

Think of AI as an incredibly efficient first-draft generator, not a finished-product creator. Your professional judgment, local expertise, ethical responsibility, and licensing obligations remain non-negotiable components of every transaction.

Throughout this guide, I'll show you exactly where AI excels and where human oversight is mandatory. I'll identify the specific checkpoints where your review isn't just recommended (it's required to protect both your clients and your career).

Real Estate AI: Opportunity or Threat?

Many agents approach AI with understandable anxiety: Will AI make my expertise obsolete? Will clients still need me? Will competitors using AI leave me behind?

These concerns echo earlier fears about online listings, virtual tours, and digital signatures (technologies that ultimately enhanced the agent's role rather than eliminating it).

Here's the truth: AI presents both opportunity and challenge.

The Opportunity: Reclaiming dozens of hours each month from mundane tasks. Hours you can reinvest in face-to-face client interactions, strategic thinking, relationship building, and the activities that truly drive your business. Or hours you can invest in yourself: your health, your family, your community, your life beyond work.

The Challenge: Learning to work alongside these tools requires an initial investment of time and willingness to experiment. Not every AI tool will work perfectly on the first try. Some will require customization to match your voice, market, and business model.

But here's what the evidence clearly shows: clients still want trusted advisors to guide them through complex transactions. They want someone who understands not just market data but the emotional journey of buying or selling a home. They want expertise, empathy, and advocacy, which are all qualities no AI system can provide.

What AI can do is amplify your capacity to deliver these uniquely human services by handling the time-consuming background work that has traditionally limited your reach.

The Strategic Balance: Human Expertise + AI Efficiency

The most successful AI implementations don't replace human expertise. They free professionals to focus on what humans do best.

In real estate, that means:

AI handles: First-draft property descriptions
You provide: Emotional appeal, local insights, strategic positioning

AI handles: Processing market data, identifying patterns
You provide: Interpretation of what patterns mean for specific clients in specific situations

AI handles: Routine follow-up sequences, appointment scheduling
You provide: Personal attention for conversations that matter, relationship building

AI handles: Document organization, deadline tracking

You provide: Strategic transaction management, problem-solving, negotiation

This human-first approach to AI adoption isn't just philosophically sound. It's practically effective. Professionals who try to automate away their expertise find themselves competing on price and efficiency alone. Those who use AI to amplify their expertise create deeper client relationships and command premium positioning in their markets.

Over thirty years of implementing complex technology solutions, I've observed a consistent pattern: the most successful technology adoptions don't replace human judgment. They eliminate friction so professionals can focus on what they do best. AI in real estate follows this same pattern.

Your First Step: Establishing The Vision That Guides Everything

Before diving into implementation details, take a moment to envision your practice and life three years from now. Agentic AI systems handle entire workflows autonomously. Your tech-resistant competitors are struggling. But you've been building toward this future systematically since applying what you read in this book.

Your new, AI-amplified typical week looks something like this:

Monday morning: Your AI systems compiled weekend showing feedback, identified patterns, and drafted personalized updates to sellers. You review, add your insights, and send. Total time: 20 minutes instead of two hours.

Tuesday afternoon: You're having coffee with a past client who's referred three buyers this year. You're not rushing. Your automated follow-up systems are nurturing your database. Your AI assistant scheduled these relationship-building meetings in the time blocks you specified.

Wednesday evening: You're at your daughter's school play. Not checking email. Not worried about missed opportunities. Your lead response system is handling inquiries. High-priority items are flagged for your morning review.

Friday: You're reviewing your week. Thirteen meaningful client interactions. Two new listings (both referrals). Three transactions moving smoothly through coordinated systems. And you logged 42 working hours instead of your old 65-hour weeks.

Saturday and Sunday: Yes, you still have showings and an open house. But instead of spending your entire weekend on paperwork and follow-up, your AI systems handle the routine tasks. You're at showings from 10am to 4pm Saturday, then done. Sunday's open house runs 1-3pm, then you have dinner with your family. The showing feedback gets automatically compiled and sent to sellers Monday morning. Follow-up emails to prospects are drafted and ready for your quick review. What used to consume your entire weekend now takes half a day, leaving real time for rest, family, and the life you're building.

This isn't fantasy. This is the practice architecture agents are building right now using the tools and strategies in this book.

The only question is: What will you do with your reclaimed time?

Will you serve more clients? Build deeper relationships? Launch that YouTube channel? Take actual vacations? Finally achieve work-life integration instead of work-life conflict?

That vision (your personal answer to "what will I do with my margin?") should guide every AI decision in this book. If an implementation doesn't move you toward that vision, skip it. If it does, prioritize it.

The Invitation

In reality, this book isn't purely about artificial intelligence as much as it is about amplified humanity.

You're about to learn dozens of specific AI applications: prompts, tools, workflows, and systems. But never lose sight of the fundamental purpose: using technology to become more of who you already are at your best.

The best agents aren't the most technical. They're the ones who use technology to reclaim their time, deepen their relationships, and focus their irreplaceable human capabilities where they matter most.

That agent can be you.

In the next chapter, I'll demystify AI terminology, set up your digital workspace, and help you complete your first implementations. The technical barriers are lower than you think. The potential returns are higher than most agents realize.

Let's begin building your amplified practice.

Chapter 2: AI Foundations for Real Estate Professionals

Before You Build: Understanding Your Tools

You wouldn't list a property without understanding its features. You wouldn't negotiate without knowing your leverage points. And you shouldn't implement AI without grasping what these tools do and how they think.

This chapter is about giving you enough understanding to use AI confidently, strategically, and effectively in your practice. Think of it as the difference between knowing how to drive a car (useful) versus knowing how to rebuild an engine (unnecessary for most people).

The concepts below will help you make better decisions about which tools to use, how to get better results, and where human oversight remains non-negotiable.

AI Terminology Demystified for Real Estate

The world of AI comes with its own vocabulary that can seem intimidating at first glance. Let me translate these terms into real estate language you can use:

Artificial Intelligence (AI): Technology that can perform tasks that typically require human intelligence. In real estate terms, think of it as a digital assistant who can help with tasks ranging from writing descriptions to analyzing market data, but one that needs your **guidance** to produce its best work.

> **Just-in-Time Learning: Why "Guidance" Matters**

AI doesn't think like humans. It recognizes patterns from massive amounts of training data. When you give it clear, specific instructions (called prompts), it produces better results. Vague requests get vague outputs. Think of it like delegating to a very literal assistant who follows instructions exactly as given.

Large Language Models (LLMs): The technology behind tools like ChatGPT, Claude, and Gemini that can generate human-like text based on the input they receive. Think of them as extremely knowledgeable assistants who need specific direction from you. Current leading models include GPT-5, Claude 4.5 Sonnet, and Gemini 3.

Generative AI: Systems that can create new content like text, images, or designs. These tools help you create listing descriptions, marketing materials, and virtually staged property photos. The most common generative AI tools you'll use are large language models like ChatGPT and Claude.

Just-in-Time Learning: What's a "Prompt"?

A prompt is simply the instruction you give an AI tool.

Prompt Engineering: The skill of crafting effective instructions for AI tools. Learning to write good prompts is like learning to delegate effectively. The clearer your instructions, the better the results. This is so important that I'll show you five detailed examples at the end of this chapter and dedicate advanced techniques in Chapter 8.

Machine Learning: A subset of AI where systems learn from data without explicit programming. For real estate, this means tools that analyze thousands of transactions to identify patterns that might predict market movements or future seller behavior.

Natural Language Processing (NLP): Technology that helps computers understand and generate human language. This powers tools that can write listing descriptions, respond to client inquiries, or summarize lengthy market reports.

AI Models: The underlying systems trained on vast amounts of data that power AI applications. Different models excel at different tasks. Some are better at writing, others at image creation or data analysis.

Algorithm: A set of rules a computer follows to solve a problem or accomplish a task. In real estate AI, algorithms might determine which properties to recommend to a client or which homeowners are likely to sell soon.

ChatGPT vs. Claude vs. Gemini: These are different AI assistants with similar capabilities but subtle differences. Think of them like having different assistants in your office with slightly different strengths:

- **ChatGPT** (GPT-5.1 and 5.2): Excellent all-around performance, strongest for creative writing and problem-solving
- **Claude** (4.5 Sonnet): Exceptional at longer, nuanced content and following complex instructions
- **Gemini** (2.5 and 3): Strong integration with Google services, good for research-heavy tasks

⚠ IMPORTANT LIMITATION: Knowledge Cutoffs

Most AI models have a "knowledge cutoff date" (the last date their training data includes). They don't automatically know current events, recent market changes, or new regulations. Always verify market data, interest rates, and regulatory information from current sources.

Just-in-Time Learning: The Human Checkpoint

Here's a fundamental truth about working with AI: **you are always the final authority**. AI generates drafts. You provide judgment, accuracy, local knowledge, and ethical oversight. Every output goes through your review before reaching clients. This isn't a limitation of AI. It's the proper architecture of amplified expertise.

Understanding Prompt Engineering: Your Most Important AI Skill

Before you set up tools or invest in software, you need to master the one skill that determines the quality of everything AI produces for you: prompt engineering.

Think of prompts as delegation. When you delegate to a team member, the quality of their work depends largely on the clarity of your instructions. AI works the same way, but with even greater literalness.

A weak prompt:

"Write a listing description"

A strong prompt:

"Create a compelling 250-word property description for a 3-bedroom Craftsman bungalow in Portland's Sellwood neighborhood. Emphasize the original 1920s woodwork, updated kitchen, and walkability to local shops and restaurants. Target buyers are likely young families or professionals seeking character homes in established neighborhoods. Use an elegant but approachable tone."

The difference in output quality between these two prompts is dramatic. The first might generate generic, forgettable copy. The second generates something you can use with minimal editing.

The Five Elements of Effective Prompts

1. Role/Context

Tell the AI what perspective to take.

Example: "You are an experienced luxury real estate agent writing for high-net-worth clients in coastal markets."

2. Task

State exactly what you want created.

Example: "Create a 300-word property description..."

3. Specifics

Provide all relevant details, constraints, and requirements.

Example: "...for a 4BR/3.5BA contemporary home with ocean views, chef's kitchen, home office, and resort-style pool. Built in 2018, 3,400 sq ft."

4. Audience

Describe who will read or use this.

Example: "Target buyers are successful professionals or early retirees seeking a primary residence with entertaining space."

5. Style/Tone

Specify how it should sound.

Example: "Use sophisticated but warm language. Avoid clichés like 'dream home' or 'stunning.' Focus on lifestyle benefits."

Not every prompt needs all five elements, but the more specific you are, the better your results.

The Iteration Mindset

Here's something most agents don't realize at first: your first prompt rarely produces perfect output. That's normal. The most effective AI users treat prompting as a conversation:

1. **Start with a solid prompt** (using the five elements above)

2. **Review the output** critically

3. **Refine your request**: "Make it more conversational" or "Focus more on the outdoor space" or "Shorten to 200 words"

4. **Iterate until satisfied**

This process that might take 5-7 exchanges when you're learning becomes 1-2 exchanges once you understand how to prompt effectively. Most agents find they're getting usable first drafts within 2-3 weeks of regular practice.

5. **Ask AI itself to make it better**

An awesome feature and benefit of working with AI tools is being able to dialogue with them. If you feel you are struggling to get the result you want from a well-crafted prompt, just ask AI to help make it better before it actually "runs it" to process and respond to. After all, AI knows the most about what it needs to arrive at your desired outcome. So ask it.

✏ Future-Ready Checkpoint: Why Prompt Mastery Matters

The agents who learn to prompt effectively now will seamlessly transition to managing more advanced and prevalent agentic AI systems in a few years from now. Those future systems won't just respond to prompts. They'll also execute complex, multi-step workflows based on high-level instructions. The skill you're building now (clear communication of intent, specific constraints, desired outcomes) directly translates to directing autonomous AI agents. Start specific, stay specific.

Five Powerful Prompts You Can Use Today

Let me give you five well-structured prompts you can copy, customize, and use immediately. Each addresses a common real estate task and includes notes on what to customize.

1. Create a Property Description (10 minutes)

> Create a compelling 250-word property description for a [PROPERTY TYPE] in [NEIGHBORHOOD/CITY].

Key details:

- [X] bedrooms, [X] bathrooms

- [X] square feet

- Built in [YEAR]

- Notable features: [LIST 5-7 KEY FEATURES]

- Location benefits: [NEARBY AMENITIES]

Target buyers are likely [DESCRIBE IDEAL BUYER PROFILE, e.g., "young families seeking good schools" or "professionals wanting walkable urban lifestyle"].

The tone should be [ELEGANT/ENTHUSIASTIC/PROFESSIONAL] and focus on [KEY SELLING POINT, e.g., "indoor-outdoor living" or "move-in ready condition"].

Avoid clichés like "dream home," "stunning," or "must-see." Use specific, vivid language that helps buyers envision their life in this home.

✎. **Customize:** Property details, target buyer profile, tone, and key selling point

⏱ **Time Saved:** 15-20 minutes per description

💧 **Human Touch Required:** Verify all facts, add hyperlocal insights, adjust for your brand voice

---

2. Draft Client Follow-Up Emails (5 minutes)

> Write a follow-up email to a buyer client who viewed three properties yesterday. They seemed most interested in the second property (the [PROPERTY TYPE] on [STREET NAME]) but expressed concerns about [SPECIFIC CONCERN].
>
> Keep the tone professional yet warm. Acknowledge their concern with a brief perspective (without dismissing it). Suggest 2-3 concrete next steps we could take. Include a clear call to action.
>
> Sign it with my name: [YOUR NAME]

✎. **Customize:** Property details, specific concern, your name

⏱ **Time Saved:** 10-15 minutes per email

💧 **Human Touch Required:** Add personal observations from the showing, adjust tone for relationship depth

---

3. Create a Week's Worth of Social Posts (15 minutes)

> Create 5 engaging real estate social media posts for [PLATFORM: Instagram/Facebook/LinkedIn].
> Include:

- 1 market update for [YOUR CITY / NEIGHBORHOOD] (use general trends, I'll add current data)

- 2 buyer or seller tips that are genuinely actionable

- 1 post highlighting my specialty in [YOUR SPECIALTY, e.g., first-time buyers, luxury homes, investment properties]

- 1 personal brand post that establishes my expertise in [YOUR NICHE] without being salesy

Each post should be 2-3 sentences with 5-7 relevant hashtags. Make them conversational and valuable, not promotional. I want to educate and build trust.

My brand voice is:
[PROFESSIONAL / APPROACHABLE / ENERGETIC / etc.]

Customize: Platform, location, specialty, niche, brand voice

Time Saved: 60-90 minutes per week

Human Touch Required: Add current market data, verify tips are locally relevant, inject personal stories

---

4. Generate Property Showing Notes Template (5 minutes)

> Create a structured template I can use to quickly record notes after showing properties to buyer clients.

Include sections for:

- Basic property information (address, MLS#, price)

- Client's immediate reactions and comments

- Specific features they liked or disliked

- Concerns or questions raised

- How this property compares to their stated priorities

- Next steps or action items

- My observations about fit

Format it so I can fill it out in 2-3 minutes right after each showing, either on my phone or laptop. Use simple bullet points and clear section headers.

✎ **Customize:** Add sections specific to your process

⏱ **Time Saved:** Creates reusable system (saves 5-10 min per showing long-term)

🖐 **Human Touch Required:** Fill in observations, client reactions, strategic notes

---

5. Analyze a Neighborhood for Clients (10 minutes)

> Help me create a neighborhood overview for [NEIGHBORHOOD NAME] in [CITY].

I need information about:

- Neighborhood character and typical resident profile

- School options (list names without making quality judgments, I'll add ratings)

- Shopping, dining, and entertainment options

- Parks and recreation facilities

- Transportation and commute options

- Recent development or notable changes

Format this as a client-friendly guide I'll supplement with current MLS data and my local expertise. Use objective, factual language. Avoid any language that could be interpreted as steering or Fair Housing violations (no references to "family-friendly," "safe," "quiet," etc.).

Keep it to 300-400 words.

✎ **Customize:** Neighborhood name, city

⏱ **Time Saved:** 30-45 minutes per neighborhood guide

🪶 **Human Touch Required:** Verify current information, add hyperlocal insights, ensure Fair Housing compliance, supplement with MLS data

⚠ **Fair Housing Alert:** Always review AI-generated neighborhood information carefully. Remove any language that references or could be interpreted as referencing protected classes. Add your local knowledge and current data. You are responsible for compliance, not the AI.

[BEGINNER ENTRY POINT] Setting Up Your Digital Workspace

If you're new to AI, here's how to create a productive digital environment. This 60-minute setup creates a foundation you'll build on throughout this book.

Step 1: Choose Your Primary AI Assistant (15 minutes)

For most real estate professionals, I recommend starting with **ChatGPT Plus** ($20/month) as your primary tool:

1. Visit chat.openai.com and create an account

2. Upgrade to ChatGPT Plus for access to GPT-5 (significantly better than the free version)

3. Bookmark this page for easy access from all your devices

4. Download the mobile app for on-the-go use

Alternative Options:

- **Claude** (free tier available at claude.ai) offers excellent performance for longer content
- **Gemini** (free tier at gemini.google.com) integrates well if you use Google Workspace

- **Perplexity Pro** ($20/month at perplexity.ai) gives you access to all three major models (ChatGPT, Claude, and Gemini) in one interface, plus real-time web search for current market data

Just-in-Time Learning: Why Pay $20/Month?

Free AI tools are excellent for testing and light use. Professional tools earn their cost through: (1) Better quality outputs that require less editing, (2) Faster response times that don't interrupt your workflow, (3) Advanced features like document analysis and web browsing, (4) Higher usage limits so you don't hit daily caps. Most agents recoup this cost by saving 30-60 minutes in the first week.

Step 2: Organize Your Workspace (10 minutes)

Create a folder on your device called "AI Resources" with subfolders:

- Successful Prompts
- Templates
- Client Communications
- Marketing Content

Set up a simple note-taking system (Google Docs, OneNote, or Notion). Create a running document called "Prompt Library" where you'll save effective prompts. This becomes your personalized playbook.

Step 3: Prepare Your Essential Data (20 minutes)

Gather examples of your best past work:

- Property descriptions that generated strong interest
- Client emails that got positive responses
- Social media posts that drove engagement

Compile information about your target neighborhoods. Write a brief description of your brand voice (professional? conversational? energetic?). List your unique value propositions and key messaging.

Why does this matter? You'll use these examples to "train" AI on your style. The more context you provide, the more your AI outputs will sound like you.

Step 4: Complete Your First Test Project (30 minutes)

Choose a low-risk task like drafting a social media post. Use one of the five prompts from the previous section. Try three variations with different specific details.

Save the most successful prompt for future use in your Prompt Library. Note what worked and what needed editing.

This low-stakes practice builds confidence before you tackle client-facing work.

Step 5: Integrate with Your Existing Workflow (15 minutes)

Identify one regular task you do weekly (newsletter, listing descriptions, client follow-ups). Develop an AI-assisted process specifically for this task.

Create a simple before-and-after comparison:

- Time required previously
- Time required with AI assistance
- Quality comparison
- Client response (if applicable)

▮ **Life Margin Impact:** If AI saves you 30 minutes on weekly newsletters, that's 26 hours annually. What would you do with an extra 26 hours? Three long weekends with family? A professional development course? Finally launching that podcast? The cumulative effect of multiple, small time savings creates significant life margin.

Remember, the goal isn't to use AI for everything immediately. Start with one or two tasks where you can see immediate benefits, then expand as your comfort grows.

Building Your AI Skills Progressively

The five prompts and setup steps above establish a foundation. In the following chapters, I'll dive deeper into more sophisticated applications and workflows.

The pattern you'll follow throughout this book:

1. Understand the capability

2. See a practical application

3. Get a tested prompt or process

4. Implement in your business

5. Measure the results

6. Refine your approach

Most agents find that AI skills develop quickly. What feels awkward in week one becomes second nature by week three. The key is consistent practice with real business needs rather than theoretical exercises.

Your Confidence Builder: AI doesn't require technical expertise. It requires clear communication and willingness to iterate. If you can delegate effectively to a human assistant, you can prompt AI effectively. The skills transfer directly.

What You've Accomplished

By completing this chapter, you've:

✓ Learned essential AI terminology in real estate context
✓ Understood how AI "thinks" and why specificity matters
✓ Mastered the five elements of effective prompts
✓ Received five detailed and specific prompts for immediate use
✓ Set up your digital workspace (if you're a beginner)
✓ Completed your first AI implementation

More importantly, you've built the foundation for everything that follows. The prompting skills you practiced here will compound as you tackle more complex applications in upcoming chapters.

Your Chapter 2 Homework

Before moving to Chapter 3, complete at least one of the five prompt examples with a real task from your business. Document your experience:

- What worked well?
- What required editing?
- How much time did you save compared to doing it manually?
- What would you do with that reclaimed time?

This isn't busywork. It's skill-building through real application. The agents and brokers who get the most value from this book are those who implement as they read, not those who read cover to cover before trying anything.

In Chapter 3, I'll help you audit your entire workflow to identify the highest-impact opportunities for AI integration. You'll learn which tasks to automate first, which to automate later, and which to keep fully human.

Let's keep building your amplified practice.

Chapter 3: Daily Workflow Integration

The Week That Opened Her Eyes

Maria had been in real estate for eight years. She knew her market cold. Her clients loved her. But she was exhausted.

One Tuesday morning, her broker suggested something unusual: "Track everything you do for one week. Every task. Every email. Every call. Just document it."

Maria resisted. She didn't have time for busy work. But she humored him and kept a simple log on her phone.

The results stunned her.

She spent 6 hours per week on transaction paperwork (most of it just organizing, formatting, and tracking). Another 4 hours creating marketing content (much of it like last month's content with different addresses). Three hours nurturing leads with follow-up emails that followed predictable patterns. Two and a half hours researching market data she could have delegated.

Nearly 16 hours per week on tasks that didn't require her expertise, local knowledge, or relationship skills. Tasks a well-instructed assistant could handle. Tasks AI could handle.

That realization changed everything. Within 30 days of implementing AI-assisted workflows, Maria reclaimed 7 hours per week. She served the same number of clients. Her service quality improved because she wasn't rushing. And for the first time in years, she made it to her daughter's Wednesday soccer practices.

This chapter shows you how to find your 7 hours (or more).

Where Do AI Tools Make a Real Difference?

You have limited hours in your day. To put AI to work effectively, start by identifying where your current workflow consumes the most time for the least reward.

Real estate agents spend significant time on tasks that don't directly generate leads or close deals: paperwork, email management, follow-ups, marketing logistics, and basic research. Some agents have shared that as many as 30 of the 40 typical hours required to close a deal are spent on administrative tasks.

The highest-impact use of AI isn't trying to automate everything overnight. It's targeting specific points of friction. By focusing on these "efficiency pressure points," you can quickly free up time to reinvest in higher-value activities.

Just-in-Time Learning: "Efficiency Pressure Points"

Efficiency pressure points are the parts of your workflow that require time without delivering a direct reward. They're usually tasks that are necessary but not unique to your expertise. These are your highest-ROI targets for AI assistance.

Common Workflow Pain Points

Most agents struggle with the same time drains:

- Writing new and unique property descriptions for every listing
- Creating and scheduling social content consistently
- Responding to routine email and text inquiries
- Drafting personalized client follow-up messages
- Crunching market data and researching comps
- Documenting feedback from showings and creating reports
- Organizing client files and transaction checklists
- Preparing market analyses and neighborhood overviews

Notice what these tasks have in common: they're necessary, they're time-consuming, and they follow patterns. That's exactly where AI excels.

🌐 **Human Touch Still Required:** These tasks benefit from AI assistance, but they still need your final review, local insights, and professional judgment. AI handles the heavy lifting. You provide the expertise and polish.

Audit Your Current Workflow

Before building an AI-enhanced workflow, you need baseline data. Spend one week tracking where your time goes. You don't need perfect precision. You need honest awareness.

In a simple spreadsheet or notebook, create these categories:

- **Client Communications** (calls, emails, texts, in-person meetings)
- **Marketing Content Creation** (listings, social posts, newsletters, videos)
- **Transaction & Paperwork** (contracts, coordination, documentation)
- **Market Analysis / Research** (CMAs, market reports, neighborhood data)
- **Lead Nurturing** (follow-ups, database management, prospecting)
- **Other Business Operations** (admin, bookkeeping, planning)

Log your hours honestly. Set a timer when you start a task. Stop it when you're done. Use a time-tracking app if that's easier. The goal is awareness, not perfection.

Sample Results from a Typical Agent:

Category	Avg Hours/Week
Transaction Paperwork	6
Marketing Content Creation	4
Lead Nurturing	3
Market Research/Analysis	2.5
Client Communications	4
Reporting/Doc Management	2
Miscellaneous/Admin	3

This agent is spending over 15 hours per week on tasks that don't directly require licensed expertise. That's nearly 40% of a 40-hour workweek spent on work that AI could assist with significantly.

Action Step: Audit Your Week

Track your time for seven days starting today. Don't change your behavior. Just document it. Pinpoint your heaviest categories and target those first for AI enhancement.

📊 **Life Margin Impact:** If you reclaim just 25% of administrative time (roughly 4 hours per week in the example above), that's 208 hours annually. That's five full 40-hour workweeks. An entire month of working hours returned to you. What would you do with an extra month every year?

Prioritizing Tasks for AI Integration

After auditing, resist the temptation to automate everything at once. That path leads to overwhelm and abandoned tools.

Instead, select **two or three workflows** with the highest cumulative hours or greatest frustration. High ROI comes from focused effort.

High-Impact Starting Points:

1. Property Descriptions

Why this matters: You write these repeatedly. They follow patterns but need customization. AI can draft in 2 minutes what takes you 20-30 minutes manually.

Potential savings: 1.5-2 hours per week

2. Client Follow-Ups & Email

Why this matters: You send dozens of similar-but-personalized emails weekly. AI can draft consistent, prompt replies you review and send in 2-3 minutes.

Potential savings: 1-2 hours per week

3. Social Media Content

Why this matters: Consistency is critical but time-consuming. Bulk-generate and edit a week's content in 20 minutes instead of spreading it across daily efforts.

Potential savings: 1.5-2 hours per week

4. Market Analysis Summaries

Why this matters: You gather data manually, then format it into client-facing reports. AI can summarize and format. You add context and insights.

Potential savings: 1-1.5 hours per week

Just-in-Time Learning: "Workflow Integration"

Integration doesn't mean replacing your judgment. It means letting AI handle the "heavy lifting" (first drafts, data processing, formatting) while you finalize and personalize. Think of AI as doing 70-80% of the work. You do the final 20-30% that requires expertise.

Sample Integration Map: Before and After

Here's what time savings look like when you implement AI-assisted workflows strategically:

Workflow Task	Old Way (Hours/Week)	AI-Enhanced (Hours/Week)	Net Savings
Listing Descriptions	2.5	0.5	2.0
Social Content	2.0	0.4	1.6
Client Emails	2.0	0.5	1.5
Market Analysis	2.0	0.8	1.2
Transaction Reports	1.5	0.3	1.2

Total hours potentially saved: 7+ per week

That's 364 hours per year. Nine full work weeks. Over two months of your life returned.

Now multiply that by the years remaining in your career. This isn't about shaving minutes off tasks. This is about fundamentally restructuring how you allocate your finite human energy.

Setting Up Your "AI Assist" Workbench

Before diving into specific workflows, create an organized system for managing AI outputs.

1. Designate an AI Workspace

Create a private workspace in Google Docs, Notion, or OneNote just for AI drafts and outputs. Keep clear folders for:

- Listing drafts
- Email templates
- Market reports
- Social posts
- Successful prompts (your Prompt Library)

Why this matters: You'll iterate on prompts and save effective ones. This workspace becomes your personalized AI playbook that improves over time.

2. Start Small

Don't try to overhaul your entire business at once. That's the path to overwhelm and abandonment.

Choose **one workflow** and run a pilot for 30 days. Compare outputs side-by-side:

- AI draft alone
- Your manual work (old way)
- Hybrid (AI draft + your editing)

Track three things:

1. **Time:** How long does each approach take?

2. **Quality:** Which produces better results?

3. **Client reactions:** Do clients notice any difference?

3. Pilot Testing Protocol

Before subscribing to expensive AI tools ($300+ per month), run a pilot in your market. Define simple metrics:

- **Time saved** compared to manual process (track actual minutes)
- **Quality improvement** based on client feedback or engagement
- **Accuracy** verified through your professional review
- **Adoption friction** (how hard is this to use consistently?)

Track these weekly for 30 days. If results don't justify the investment, re-evaluate or pivot to different tools.

ROI Reality Check: A tool that costs $300/month needs to save you roughly 5-6 billable hours per month to break even (assuming a $50-60/hour value on your time). Most effective AI tools save that in the first week.

[INTERMEDIATE ENTRY POINT] Building a Hybrid AI Workflow

For agents comfortable with basic AI tools (from Chapter 2), here's a complete workflow for property marketing that integrates AI at multiple touchpoints.

Step 1: Draft Listing Description (AI First Pass)

Prompt:

> Create a 200-word property description for a [3-bedroom, 2-bath] ranch home in [South Tampa].
>
> Highlight these key features:
>
> - Solar panels (recent installation, utility savings)
>
> - Updated kitchen with quartz counters
>
> - Mature oak trees providing shade

- 10-minute commute to downtown Tampa

Target buyers are likely young families or professionals seeking energy efficiency and convenience.

Tone: Professional but warm. Avoid clichés. Focus on lifestyle benefits.

⏱ **Time Required:** 2 minutes to customize and run prompt

Your Value Add: Review output. Add unique seller stories, local restaurant recommendations, or specific market positioning.

Step 2: Personalize Output (5 minutes)

Read the AI draft carefully. Enhance it with:

- Hyperlocal details AI can't know (the neighborhood's Sunday farmers market, the street's annual block party)
- Recent comparable sales for context
- Unique property story if the sellers shared one
- Your strategic positioning for current market conditions

Step 3: Generate Social Posts (3 minutes)

Prompt:

> Generate five Instagram posts for this property, each highlighting a different feature:

Property: 3BR/2BA ranch in South Tampa with solar panels, updated kitchen, mature landscaping, quick downtown access

For each post:

- Write 2-3 engaging sentences

- Include 5-7 relevant hashtags

- Suggest which photo to feature (describe the shot)

- Vary the angle (cost savings, lifestyle, location, design, investment value)

Keep the tone energetic but professional.

⏱ **Time Required:** 3 minutes to generate, 10 minutes to edit and customize

Old Way Time: 45-60 minutes to create five unique posts from scratch

Step 4: Email Client Update (4 minutes)

Prompt:

> Draft a weekly email update on transaction status for my seller client.

Current status:

- Property listed 6 days ago

- 12 showings scheduled

- 2 second showings requested

- Positive feedback on kitchen and outdoor space

- One concern about lot size (smaller than expected)

Next steps:

- Broker open house this Thursday

- Evaluate offers by Friday EOD if we receive any

- Consider small price adjustment if no offers after 14 days

Tone: Warm, professional, confident. Reassure them about the strong showing activity.

Sign it: [Your name]

⏱ **Time Required:** 2 minutes to customize prompt, 2 minutes to personalize output

Old Way Time: 15-20 minutes to compose from scratch

Step 5: Create Market Snapshot (5 minutes)

Prompt:

> > Summarize current market trends for [South Tampa residential properties] based on this data:
>
> [Paste recent MLS statistics: median price, days on market, inventory levels, price per square foot]
>
> Create a client-friendly summary covering:
>
> - Overall market direction (buyer's/seller's/balanced)
>
> - Price trends compared to last quarter
>
> - Inventory levels
>
> - What this means for sellers timing the market now
>
> Keep it to 150-200 words. Use clear language, avoid jargon.

⏱ **Time Required:** 3 minutes to gather data and run prompt, 2 minutes to verify accuracy

Old Way Time: 20-30 minutes to research, analyze, and write

Step 6: Verification Checkpoint (Always)

Before any AI-generated content reaches clients:

✓ Verify all facts and figures

✓ Check for Fair Housing compliance

✓ Ensure it matches your brand voice

✓ Add hyperlocal insights AI can't provide

✓ Confirm it serves the client's specific needs

⚠ **Professional Responsibility:** You remain the licensed professional of record. AI is your drafting assistant, not your replacement. This checkpoint is non-negotiable.

Real Agent Story: Workflow Integration in Action

Amy Stockberger, Broker

The Challenge: Running a brokerage while maintaining high service standards consumed 60+ hours per week. Administrative tasks and content creation left little time for strategic growth or agent support.

The Implementation: Amy systematically automated 60% of her business operations using AI. She implemented AI-assisted workflows for listing content, market reports, client communications, and transaction coordination.

The Results: Thousands of hours saved annually while maintaining (and in many cases improving) service quality. The key was combining AI efficiency with human expertise at critical touchpoints.

The Life Impact: Time reclaimed allowed her to focus on high-value activities: agent coaching, strategic planning, and market positioning. More importantly, she rebuilt work-life boundaries that had disappeared.

Source: Inman, "How To Automate 60% Of Workflow Without Sacrificing Service", June 12, 2025. https://www.inman.com/2025/06/12/how-to-automate-60-of-workflow-without-sacrificing-service/

What This Teaches Us: The agents seeing dramatic results aren't using AI to replace their expertise. They're using it to eliminate the administrative friction that prevented them from fully deploying that expertise.

Measuring Impact: Small Steps, Big Results

After one month implementing AI-enhanced workflows, conduct a formal review:

Week 4 Assessment:

Time Saved:

- Audit how much time you've saved in each category you targeted
- Calculate total weekly savings
- Multiply by 52 to understand annual impact

Quality Improvements:

- Note improvements in marketing reach (views, engagement, inquiries)
- Track client response rates to your communications
- Monitor deal cycle times (are transactions moving faster?)

Tool Performance:

- Document which tools and prompts deliver consistently high quality
- Identify which require the most editing or correction
- Note any tools that aren't earning their cost

Workflow Refinements:

- Revise your approach based on what worked and what didn't
- Add backup workflows for when AI outputs miss the mark
- Scale successful approaches to additional areas

Life Margin Calculation:

If you saved 6 hours per week over 30 days, that's 24 hours in one month. Project that annually: 312 hours (nearly 8 full work weeks).

What did you do with that time?

- Took on 3 more clients without working more hours?
- Finally launched that YouTube channel?

- Made it to every family dinner that month?
- Felt less stressed and more present?

Document these qualitative improvements. They matter as much as the quantitative data.

Future-Ready Checkpoint: Building Systematic Thinking

The workflow audit and integration process you're learning now (identify bottlenecks, test solutions, measure results, iterate) is exactly the skillset required to manage agentic AI systems. Those systems will require you to define desired outcomes, set constraints, and evaluate performance. You're building that muscle now with simpler tools. You'll be ready to hand entire workflows to autonomous systems because you understand how to architect, measure, and refine processes.

Troubleshooting: Common Pitfalls to Avoid

Pitfall 1: Skipping the Review Step

AI-generated content is only as good as your oversight. It can miss local nuance, include outdated information, or generate content that sounds good but isn't accurate. Never send drafts directly to clients without review.

Solution: Build the review step into your workflow as non-negotiable. Set a 5-minute timer if needed. Make it a habit, not an option.

Pitfall 2: Over-Automation

Don't automate away relationship-building. Use AI to free up time for more personal touches, not fewer. The goal is amplification, not replacement.

Solution: For every hour AI saves you, allocate at least 30 minutes to high-value human activities: client calls, handwritten notes, coffee meetings, community involvement.

Pitfall 3: Investing Too Quickly

Expensive tools promise amazing results. Some deliver. Some don't. Don't commit to annual contracts without thorough testing in your specific market.

Solution: Always run a 30-day pilot using the protocol outlined earlier. Verify value before investing heavily.

Pitfall 4: Ignoring Knowledge Cutoffs

AI models don't automatically know today's interest rates, new regulations, or hyperlocal trends. They're working from training data with a cutoff date.

Solution: Always verify time-sensitive information (rates, regulations, current inventory) from authoritative sources. Use AI for drafting and analysis, not for real-time data.

Pitfall 5: Generic Implementation

Copying someone else's workflow without adapting it to your market, brand, and clients leads to mediocre results.

Solution: Use the examples in this book as starting points. Customize everything. Test variations. Build your unique system.

Your Next Step: The One-Week Integration Challenge

Pick **one workflow** from the examples above to pilot over the next seven days:

Days 1-2: Set up your AI workspace and gather necessary data (past examples, property details, market stats)

Days 3-5: Implement the workflow. Document time spent and results achieved.

Days 6-7: Compare results to your old manual process. Calculate time saved. Note quality differences.

Track these metrics:

- Actual minutes saved per use
- Quality of output (scale of 1-10 compared to manual)
- Amount of editing required
- Client or colleague feedback (if applicable)

By day 7, you'll have concrete data showing whether this workflow is worth scaling or needs adjustment.

◎ **Success Metric:** If your chosen workflow saves you even 20 minutes per week, that's 17+ hours per year. One workflow. Now imagine implementing 4-5 of these.

What You've Accomplished

By completing this chapter, you've:

✔ Audited your current workflow to identify efficiency pressure points
✔ Prioritized high-impact tasks for AI integration
✔ Learned how to set up an organized AI workspace
✔ Seen a complete hybrid workflow from start to finish
✔ Studied a real agent's transformation story
✔ Received a framework for measuring results
✔ Identified common pitfalls and how to avoid them

More importantly, you've started building the systematic thinking that will serve you for years as AI capabilities evolve.

In Chapter 4, I'll guide you through selecting, vetting, and validating specific AI tools for your business. You'll learn how to evaluate tools, avoid expensive mistakes, and build a cost-effective AI toolkit that matches your budget and business model.

Let's keep building your amplified practice.

Chapter 4: Core AI Tools for Real Estate Agents

The $300 Mistake

David had just closed his best quarter ever. Twelve transactions, stellar client reviews, momentum building. He was ready to level up his tech stack.

He saw demos for three different "AI-powered real estate platforms." Each promised to revolutionize his business. Each cost $300-500 per month. The sales presentations were slick. The features looked incredible. He signed up for all three.

Three months later, he was using one sporadically and had forgotten the passwords to the other two. He'd spent $2,700 with minimal return. The tools weren't bad. They just didn't match his actual workflow, his market, or his current business needs.

David's mistake wasn't investing in AI tools. It was investing without a framework for evaluation.

This chapter gives you that framework. You'll learn how to evaluate tools strategically, avoid expensive mistakes, and build a cost-effective AI stack that serves your business model and market reality.

The Tool Selection Paradox

Here's the challenge facing agents: you have access to more powerful AI tools than ever before, at lower prices than seemed possible two years ago. But abundance creates its own problem.

When you search "AI tools for real estate," you'll find 50+ options. Many look similar. All claim impressive capabilities. Most have free trials that expire before you've truly tested them. How do you choose?

The answer isn't finding the "best" tools (there's no universal best). The answer is finding the right tools for:

- Your specific business model (solo agent vs. team vs. broker)
- Your current volume and growth trajectory
- Your budget and ROI requirements
- Your technical comfort level
- Your market's specific characteristics

The Guiding Principle: Always aim for proven, high-output tools at the lowest viable cost. Cheaper in price, not in quality. Maximum value per dollar spent.

The Five-Factor Tool Evaluation Framework

Before subscribing to any AI tool, evaluate it across five dimensions:

1. Problem-Solution Fit

Ask: Does this tool solve an actual problem I have right now, or a hypothetical future problem?

Tools marketed on potential future value rarely deliver. Tools that address current pain points almost always do.

Red flag: "This could be useful someday"

Green light: "This solves the problem costing me 4 hours every week"

2. Output Quality vs. Effort Required

Ask: Does this tool produce outputs I can use with 5-10 minutes of editing, or does it require 30+ minutes of cleanup?

The value proposition of AI is reducing work, not creating different work. If a tool requires extensive correction, it's not saving you time.

Test: Use the tool for the same task three times. If quality doesn't improve by the third use, the tool isn't a fit for your needs.

3. Integration with Existing Workflow

Ask: Does this tool connect with the systems I already use, or does it create a separate silo?

Tools that require you to export data, manually transfer information, or maintain duplicate records create friction. Friction kills adoption.

Green light: Direct integration with your CRM, transaction management platform, or MLS

Yellow flag: Manual export/import required

Red flag: Completely separate system with no export capability

4. True Cost of Ownership

Ask: What's the total cost including subscription, learning curve, and maintenance?

A $20/month tool that takes 2 hours to learn and works immediately has a lower true cost than a $50/month tool that requires 10 hours of setup and troubleshooting.

Calculate: (Monthly subscription × 12) + (Setup time × your hourly value) + (Monthly maintenance time × your hourly value × 12)

5. Pilot Performance

Ask: In a 30-day pilot, did this tool deliver measurable improvement?

Never commit to annual contracts without running a pilot. Monthly subscriptions let you test with minimal risk.

Track: Time saved, quality improvement, client response (if applicable), ease of use

⌨ Future-Ready Checkpoint: Building Selection Discipline

The tool evaluation discipline you're building now becomes even more critical as AI capabilities expand. A few years from now, you'll have hundreds more options, many with agentic capabilities. Agents who've developed systematic evaluation criteria won't be overwhelmed by the vast assortment of AI agents. Rather, they'll quickly identify which new capabilities are worth piloting. Those who chase every new tool will drown in subscriptions and complexity.

The Three Tool Categories

AI tools for real estate fall into three broad categories. Understanding these categories helps you build a balanced stack.

Category 1: General-Purpose AI Assistants

These are your "Swiss Army knife" solutions. Flexible, powerful, increasingly capable across text, images, documents, and research.

When to use them: Daily communication, content creation, analysis, brainstorming, research

What they can't do: Access your specific CRM data, integrate with your MLS, or understand your local market nuances without your input

Current leaders (as of December 2025):

- **ChatGPT** (GPT-5.1, GPT-5.2)
- **Claude** (4.5 Sonnet)
- **Gemini** (2.5, 3)
- **Perplexity Pro** (provides access to all three above plus real-time web search)

Pricing reality: $20-30/month for professional tiers

You've already been introduced to these in Chapter 2, so I won't repeat capability descriptions here. The key question for Chapter 4 is: which one should be your primary tool?

Decision Framework:

Choose ChatGPT if:

- You want the most versatile all-around performance
- You value extensive third-party integrations
- You need strong creative and problem-solving capabilities

Choose Claude if:

- You work with longer documents frequently
- You need nuanced, detailed content
- You appreciate more "thoughtful" responses

Choose Gemini if:

- You live in Google Workspace (Docs, Sheets, Gmail)
- You need strong research and web integration
- You want tight integration with Google services

Choose Perplexity Pro if:

- You want access to all three models in one interface
- You need real-time web search for current market data
- You value having options without multiple subscriptions

My recommendation for most agents: Start with either ChatGPT Plus or Perplexity Pro. Both offer exceptional value at $20/month. Add a second tool only if you identify a specific gap after 60 days of use.

Category 2: Real Estate-Specific AI Tools

These tools are designed for industry-specific tasks. They often deliver better results for their specific use case than general-purpose tools.

When to use them: Property marketing (staging, enhancement), lead qualification, predictive analytics, compliance-focused content

What they can't do: Adapt to uses outside their designed purpose

Let me give you the current landscape with accurate pricing (as of December 2025):

Virtual Staging & Photo Enhancement

Virtual Staging AI

Website: virtualstagingai.app

Best for: Creating realistic virtual staging for empty rooms

Current Pricing:

- Basic: $16/month (6 photos/month) = $2.67/photo
- Standard: $19/month (20 photos/month) = $0.95/photo
- Professional: $39/month (unlimited renders)
- Enterprise: $79/month (unlimited + API access)

Sweet spot: Standard plan at $19/month for most agents

Output quality: Very good to excellent for standard residential spaces

ROI calculation: Traditional staging costs $2,000-5,000. Virtual staging at $19/month covers all your listings for less than the cost of one physically staged property.

ReimagineHome

Website: reimaginehome.ai

Best for: Photo enhancement, virtual staging, exterior modifications

Current Pricing:

- Free tier: 5 generation credits (testing only)
- Legacy Tools: $14/month (30 credits first month, then 20/month) at $0.47/credit
- Full Design Studio: $22/month (200 credits first month, then 150/month) at $0.15/credit
- Power: $40/month (400 credits/month) at $0.10/credit
- Agency: $80/month (1000 credits/month) at $0.08/credit

Sweet spot: Full Design Studio at $22/month for comprehensive capabilities

Output quality: Good to excellent, especially for exterior enhancements

Key difference from Virtual Staging AI: More versatile (interior and exterior), credit-based system allows flexibility

Lead Management & Client Engagement

Lofty AI Assistant

Website: lofty.com/ai/sales_assistant

Best for: Automated lead qualification, nurturing, and appointment setting

Current Pricing:

- Not transparently published (requires quote)
- Core CRM: ~$449-700/month (based on third-party reviews)
- AI Assistant add-on: $99/month (confirmed from partnership documentation)
- Total starting cost: ~$548-799/month minimum

Output quality: Good for lead engagement and qualification

Reality check: This is enterprise pricing. Solo agents should explore this only after closing 2+ transactions monthly. Teams of 3+ agents can justify the cost through distributed value.

What it does: Virtual ISA that captures leads, matches properties, engages in conversations, and handles agent handoffs. Effective for lead nurturing and appointment setting with minimal agent involvement.

Top Producer Smart Targeting

Website: topproducer.com/smart-targeting

Best for: Predictive analytics to identify potential sellers

Current Pricing:

- Pro + Farming (includes Smart Targeting): $399/month per user
- Includes full CRM functionality plus AI-powered predictive targeting

Output quality: Good predictive accuracy for identifying seller probability

What it does: Analyzes 2000+ attributes and 20 years of listing data to predict which homeowners in your real estate farm area are most likely to sell in the next 12 months. Includes multi-channel marketing automation (Facebook ads, postcards, CMAs, emails).

ROI threshold: You need to close one additional transaction every 4-5 months from these leads to justify the cost.

Category 3: AI-Enhanced Traditional Tools

Many tools you already use now include AI features. Before adding specialized AI tools, audit what you already have access to.

CRMs with AI Components:

- Follow Up Boss (AI-assisted task prioritization, smart lead scoring)
- Real Geeks (AI lead response, automated nurturing)
- Lofty (comprehensive AI features at various pricing tiers)

Marketing Platforms with AI:

- **Canva Pro** ($10/month or $120/year): AI-powered design, Magic Write for copy, background removal, image enhancement
- Adobe Express: AI design features included with Creative Cloud subscriptions

Transaction Management Platforms:

- Dotloop: AI-assisted document completion, automated reminders
- Skyslope: Risk alerts, deadline tracking with intelligent notifications

Market Analysis Tools:

- CoreLogic: Advanced analytics based on massive transaction datasets
- HouseCanary: Predictive market modeling
- Redfin Data Center: Market trend analysis with AI-enhanced insights

Before subscribing to new tools, maximize what's already in your existing stack. In a hurry, many agents pay for AI features they never activate in tools they already own.

Building Your AI Stack: Three Tiers for Different Agent Types

Not every agent needs the same toolkit. Your ideal stack depends on your business model, volume, and growth stage.

Tier 1: The Foundation Agent

Profile: Solo agent, 1-10 transactions/year, building volume

Budget: $20-50/month

Primary Goal: Time reclamation on repetitive tasks

Recommended Stack:

- **ChatGPT Plus or Perplexity Pro** ($20/month): Primary assistant for all text-based tasks
- **Canva Pro** ($10/month): Marketing visuals, social content, property flyers
- **Your CRM's built-in AI features** (no additional cost)

Total monthly cost: $30-50

Expected outcome: Reclaim 5-10 hours/week, improve marketing consistency

What you don't need yet: Specialized real estate tools, predictive analytics, enterprise CRM features

---

Tier 2: The Scaling Agent

Profile: Solo or small team, 10-30 transactions/year, consistent volume

Budget: $50-150/month

Primary Goal: Systematic lead nurturing and marketing excellence

Recommended Stack:

- **ChatGPT Plus or Perplexity Pro** ($20/month)
- **Canva Pro** ($10/month)
- **Virtual Staging AI Standard or ReimagineHome Full Studio** ($19-22/month)
- **Your CRM's AI features** (maximize what you're already paying for)
- **Consider adding:** Claude Pro for long-form content ($20/month)

Total monthly cost: $50-72 (or $70-92 with Claude)

Expected outcome: 30%+ productivity increase, serve 2x clients with same stress level, professional-grade marketing

What you don't need yet: Enterprise predictive tools, multiple redundant AI assistants

---

Tier 3: The Amplified Team Leader

Profile: Team lead or high-volume solo agent, 30+ transactions/year

Budget: $150-500/month

Primary Goal: Full workflow automation, competitive intelligence, team leverage

Recommended Stack:

- **Multiple general-purpose AI tools** (ChatGPT, Claude, Perplexity for different team members and use cases)
- **Professional photo tools** (Virtual Staging AI Professional or ReimagineHome Agency tier)
- **Canva Pro** (team account)
- **Consider:** Top Producer Smart Targeting ($399/month) if farming specific areas
- **Consider:** Lofty AI Assistant ($99+ add-on) for team lead management
- **Advanced CRM features** fully activated and optimized

Total monthly cost: $150-500+

Expected outcome: 3x client capacity, true "CEO of practice" positioning, predictable lead flow

ROI requirement: At this level, every tool must demonstrably contribute to closed transactions. Track rigorously.

📊 **Life Margin Impact by Tier:**

- **Tier 1:** Reclaim 5-10 hours/week (260-520 hours/year) for $360-600 annual investment
- **Tier 2:** Reclaim 10-15 hours/week (520-780 hours/year) for $600-1,104 annual investment
- **Tier 3:** Reclaim 15-20 hours/week (780-1,040 hours/year) while serving 2-3x more clients

What would you do with an extra 500-1,000 hours per year?

The 30-Day Pilot Protocol

Never commit to expensive tools without systematic testing. Here's the exact protocol to follow:

Week 1: Baseline Measurement

Before activating any new tool, document your current state:

1. **Select 2-3 specific tasks** the tool is supposed to improve

2. **Track time spent** on these tasks for one full week (use a timer, be honest)

3. **Assess current quality** using a simple 1-10 scale

4. **Note frustration points** where the task feels most tedious

Example baseline:

- Task: Creating listing descriptions
- Current time: 25 minutes average per listing
- Current quality: 7/10 (effective but formulaic)

- Frustration point: Finding fresh language for similar properties

Week 2-3: Active Testing

1. **Activate the tool** (use monthly billing, not annual)

2. **Use it for the same tasks** you measured in Week 1

3. **Track actual time spent** including learning curve and editing

4. **Assess output quality** using the same 1-10 scale

5. **Note unexpected benefits or drawbacks**

Important: Don't change your workflow dramatically. Test the tool in your actual work environment.

Week 4: Analysis & Decision

Calculate three metrics:

1. Time ROI:

Old process time - New process time = Time saved per use

Time saved per use × Frequency per month = Monthly time savings

Monthly time savings × Your hourly value = Monthly dollar value

2. Quality Impact:

New quality score - Old quality score = Quality change

If quality decreased, the tool fails regardless of time savings

3. Total Cost:

Monthly subscription + (Learning time × Your hourly value) +

(Any integration costs) = True monthly cost

Decision Framework:

Keep the tool if:

- Monthly dollar value $\geq$ 3x true monthly cost (300% ROI minimum)
- Quality maintained or improved
- You used it consistently for 3+ weeks

Revisit in 60 days if:

- ROI is 150-299%
- Quality is equal but not better
- You see potential but adoption was inconsistent

Cancel if:

- ROI is under 150%
- Quality decreased
- You used it fewer than 10 times in 30 days
- Learning curve feels excessive for the benefit

Pilot Template

I've created a simple spreadsheet template for tracking pilots. You'll find it in Appendix B: Tool Comparison Matrix. Use it for every tool over $35/month before committing to longer contracts.

Avoiding the Tool Trap: Five Common Mistakes

Mistake 1: Buying Before Trying

The trap: Tool looks great in demo. You sign up for annual plan to save 20%. Three months later, you're not using it.

The fix: Always start monthly. The 20% annual savings aren't worth the risk of paying for a full year of a tool you abandon in month two.

Mistake 2: Confusing Features with Value

The trap: Tool has 47 features. Your competitor mentioned it. You assume more features = better tool.

The fix: You only need 3-5 features to work excellently. Judge tools by how well they solve your specific problems, not by feature count.

Mistake 3: Ignoring Integration Costs

The trap: Tool costs $50/month (reasonable). But it requires 4 hours of setup, doesn't integrate with your CRM, and creates duplicate data entry.

The fix: Calculate true cost including setup time, learning curve, and workflow friction. A $20/month tool that works seamlessly can deliver higher value than a $50/month tool that requires constant maintenance.

Mistake 4: Subscribing to Redundant Tools

The trap: You have ChatGPT Plus, Claude Pro, and Gemini Advanced because you tried each one. Now you're paying $60/month for three tools that do 80% of the same things.

The fix: Pick one primary general-purpose assistant. Add a second only if you've identified a specific, recurring gap the first can't fill.

Mistake 5: Chasing the Newest Tool

The trap: Every week a new "game-changing" AI tool launches. You try them all. You master none.

The fix: Evaluate new tools quarterly, not weekly. Build mastery with current tools before adding new ones. Depth beats breadth.

Keeping Your Stack Fresh Without Overload

AI tools evolve rapidly. Models improve. Features expand. Pricing changes. How do you stay current without constant disruption?

Monthly: Quick Check

- Scan your primary tool's changelog or updates blog (5 minutes)
- Note any new features that might solve existing problems
- Test one new feature if relevant

Quarterly: Strategic Review

- Audit your actual usage of each paid tool (do you really use it?)
- Check for new competitors or alternatives
- Reassess whether your current stack still matches your business stage
- Cancel tools you haven't used in 60+ days

Annually: Major Evaluation

- Complete fresh tool evaluation for your primary stack
- Consider whether your tier (Foundation/Scaling/Amplified) has changed
- Look for emerging tool categories that didn't exist last year
- Renegotiate pricing or switch to annual billing only for proven tools

Trusted Information Sources

Stay informed without drowning in hype:

- **Inman** (inman.com): Real estate tech news, agent perspectives
- **HousingWire** (housingwire.com): Industry trends, tech coverage
- **T3 Sixty** (t3sixty.com): Real estate technology research

- **Your actual peers**: Ask agents in your market what they're using

Filter noise: If you hear about a tool from three trusted sources and it solves a problem you have, investigate. If you hear about it from marketing emails only, ignore it.

The Data Portability Principle

One final critical consideration: never get locked into proprietary systems that won't export your data.

Before committing to any tool, verify:

- Can I export my data in standard formats (CSV, PDF)?
- If I cancel, do I retain access to my historical data?
- Can I transfer my data to competing tools if needed?

Tools that trap your data are making a bet that switching costs will keep you paying even if value declines. Don't accept that bet.

Best practice: Every quarter, export your data from critical systems. Store it securely. You should be able to walk away from any tool with all your data intact.

Your Action Plan: Building Your Stack This Week

Day 1-2: Assessment

1. List your top 5 workflow frustrations

2. Calculate your current monthly tool costs

3. Identify which tier (Foundation/Scaling/Amplified) matches your business

4. Note what AI features you already have in existing tools

Day 3-4: Research

1. For each frustration, identify 2-3 potential tools

2. Check pricing and integration capabilities

3. Read recent reviews from real estate professionals

4. Shortlist 1-2 tools to pilot

Day 5-7: First Pilot

1. Sign up for monthly billing on your top choice

2. Begin your 30-day pilot protocol

3. Document your baseline before using the new tool

4. Commit to using it at least 3x per week

Week 2-4: Evaluate and Decide

1. Complete your pilot metrics

2. Make keep/cancel decision

3. If keeping, optimize your use

4. If canceling, try your second choice or reassess the problem

---

What You've Accomplished

By completing this chapter, you've:

✓ Learned the five-factor framework for evaluating any AI tool

✓ Understood the three tool categories and when to use each

✓ Received accurate, current pricing for major real estate AI tools

✓ Identified which tier matches your business stage

✓ Learned the 30-day pilot protocol for testing tools systematically

✓ Discovered the five most common tool selection mistakes and how to avoid them

✓ Built a strategy for staying current without chasing every new release

More importantly, you've developed systematic judgment about tool selection. This framework works for evaluating tools that don't exist yet. As AI capabilities expand, your evaluation discipline ensures you'll adopt strategically, not reactively.

Transition to Chapter 5

You've built your evaluation framework. You've identified your tier. You've selected your initial stack (or you're mid-pilot on your first tools).

Now comes the most valuable part of this book: learning exactly how to deploy these tools across every aspect of your real estate practice.

Chapter 5 provides detailed, step-by-step workflows for property marketing, client communications, market analysis, transaction management, and business development. Each section includes tested prompts, time savings estimates, and clear markers for where human oversight remains essential.

You've chosen your tools wisely. Now I'll show you how to use them masterfully.

Chapter 5: Task-Specific Applications

The Transformation Jennifer Didn't See Coming

Jennifer had been using ChatGPT for three months. She'd written maybe a dozen property descriptions. Created a few social posts. It was helpful, she supposed, but not revolutionary.

Then one Thursday afternoon, facing a weekend with six listings to prepare, she decided to follow a systematic process. She opened the prompt templates from Chapter 2. She blocked 90 minutes. She worked through each listing methodically.

By 6:15 PM, she had:

- Six complete property descriptions (edited and polished)
- Thirty social media posts scheduled across two weeks
- Personalized email updates drafted for four active clients
- A market analysis compiled for a Monday listing presentation

Tasks that normally consumed her entire Friday and half of Saturday were done. She had her weekend back.

But here's what surprised her most: the quality was better. Not just faster, better. Because she wasn't rushing, she could add the local insights and personal touches that made her marketing distinctive. She had time to think strategically about positioning, not just get words on paper.

That's when Jennifer understood: **AI doesn't just save time. It creates the space for you to do your best work.**

This chapter shows you exactly how to create those moments. Not theoretical possibilities, but detailed, step-by-step workflows for every major task in your real estate practice.

How to Use This Chapter

This is not a chapter to read straight through. It's a reference guide organized by business function.

Five major sections:

1. **Property Listings & Marketing** (descriptions, photos, videos)

2. **Client Communications** (emails, social media, relationship management)

3. **Property Research & Analysis** (CMAs, market reports, neighborhood guides)

4. **Transaction Management** (documents, timelines, coordination)

5. **Business Development** (lead qualification, time management, productivity)

For each workflow, you'll find:

- ⏱ **Time Saved:** How much time this typically reclaims

- ☺ **Expertise Required:** Difficulty level (●●○○○ = beginner friendly)

- 🌑 **Human Touch Required:** How much personalization you need to add

- ⅈ **Life Margin Impact:** What you could do with the reclaimed time

- ⚠ **Checkpoints:** Where your professional review is mandatory

Your Implementation Strategy:

Don't try to implement everything at once. Instead:

1. **Scan the five sections** and identify your biggest time drain

2. **Start with one workflow** in that section

3. **Use it 5-10 times** until it feels natural

4. **Document your time savings** and quality improvements

5. **Add a second workflow** only after mastering the first

By the end of 60 days following this approach, you'll have 4-6 optimized workflows saving you 8-12 hours weekly. That's the path to sustainable transformation.

Access or download all these prompts so you don't need to retype them...

https://go.howdoiai.pro/re-all-prompts-e1

Let's begin.

Section 1: Property Listings & Marketing

Creating Compelling Property Descriptions

The Reality: You write 2-3 property descriptions per month (if you're active, maybe 5-8). Each takes 20-30 minutes. Most sound similar because you're trying to describe different houses using the same vocabulary.

The Opportunity: AI can generate a solid first draft in 90 seconds. You spend 5-7 minutes adding your expertise, local insights, and strategic positioning. Result: better descriptions in one-fourth the time.

⏱ **Time Saved:** 15-20 minutes per listing

◎ **Expertise Required:** ●●○○○ (Beginner friendly)

🖱 **Human Touch Required:** ●●●○○ (Medium, add local knowledge and verify accuracy)

The Four-Step Process

Step 1: Gather Your Intelligence (3 minutes)

Before prompting, compile:

- Basic specs (beds, baths, square footage, year built)
- 5-7 standout features (what makes this property unique?)
- Neighborhood assets (what's nearby that buyers care about?)
- Target buyer profile (who's this perfect for?)
- Your strategic angle (what's the key selling message?)

Pro tip: Create a simple checklist on your phone. Fill it out during your listing walkthrough. Having this ready makes the AI prompt faster and better.

Step 2: Use the Tested Prompt Template (2 minutes)

> Create a compelling [WORD COUNT]-word property description for a [PROPERTY TYPE] in [NEIGHBORHOOD/CITY].

Key details:

- [X] bedrooms, [X] bathrooms

- [X] square feet

- Built in [YEAR]

- Notable features: [LIST 5-7 KEY FEATURES]

- Location benefits: [NEARBY AMENITIES]

Target buyers are likely [DESCRIBE IDEAL BUYER PROFILE, e.g., "young families seeking good schools and walkability" or "professionals wanting low-maintenance luxury near downtown"].

The tone should be [ELEGANT/ENTHUSIASTIC/PROFESSIONAL/SOPHISTICATE

> D] and focus on [KEY SELLING POINT, e.g., "indoor-outdoor living" or "move-in ready condition" or "investment potential"].
>
> Avoid real estate clichés like "dream home," "stunning," "must-see," or "won't last long." Use specific, vivid language that helps buyers envision their life in this home, not generic superlatives.

Why this works: Specificity. Vague prompts get vague outputs. This prompt gives AI everything it needs: facts, audience, angle, tone, and constraints.

Step 3: Review and Enhance (5-7 minutes)

AI gave you structure and language. Now add what AI can't know:

☑ **Verify every fact** (AI can't access your MLS, it's working from your prompt)

☑ **Add hyperlocal details** ("two blocks from Riverside Elementary's award-winning STEAM program")

☑ **Include market positioning** ("priced $15K below recent comparables for quick sale")

☑ **Inject personality** (your voice, your brand, your strategic angle)

☑ **Check Fair Housing compliance** (remove any language referencing protected classes)

Step 4: A/B Test Your Descriptions (ongoing)

Over time, track which description styles generate the most:

- Online views and saves
- Showing requests
- Buyer inquiries

Refine your prompt template based on what works in your market.

▌ Life Margin Impact:

If you write 4 listings per month and save 18 minutes each, that's 72 minutes monthly or 14.4 hours annually. What would you do with an extra 14 hours? Attend three more of your kid's games? Finally read that professional development book? Take a long weekend without work stress?

Sample Prompt: Family Home in Suburban Market

> Create a compelling 250-word property description for a 4-bedroom colonial home in Naperville, Illinois.

Key details:

- 4 bedrooms, 2.5 bathrooms

- 2,400 square feet

- Built in 1998, extensively updated in 2023

- Notable features: Renovated kitchen with quartz counters and stainless steel appliances, hardwood floors throughout main level, finished basement with rec room and home office nook, large backyard with mature oak trees and professional landscaping, two-car garage

- Location benefits: Walking distance to Madison Elementary (rated 9/10), three blocks from Prairie Path bike trail, 5-minute drive to downtown Naperville shops and restaurants

Target buyers are likely families with school-age children seeking space, convenience, and an established neighborhood with strong community feel.

The tone should be warm and family-focused, and focus on the lifestyle this home enables for growing families.

Avoid real estate clichés like "dream home," "stunning," "must-see," or "won't last long." Use specific, vivid language that helps buyers envision their family life in this home, not generic superlatives.

⚠ PROFESSIONAL RESPONSIBILITY CHECKPOINT:

You remain responsible for every word in published descriptions. Verify facts, ensure Fair Housing compliance, and confirm all statements are accurate and defensible. AI drafts, you approve.

Future-Ready Checkpoint:

The description-writing skill you're building now (clear inputs, strategic framing, quality review) directly translates to managing agentic AI in the years to come. Future systems will generate complete marketing packages (description, social posts, email campaigns, ads) from a single comprehensive brief. Agents who master strategic prompting now will seamlessly upgrade to directing autonomous marketing systems.

Enhancing Property Photos with Virtual Staging

The Reality: Physical staging costs $2,000-5,000 per property. Not every seller can afford it. Empty rooms photograph poorly and depress perceived value.

The Opportunity: Virtual staging costs $0.95-2.67 per photo (based on Chapter 4 pricing). Professional quality. Multiple style options. Done in minutes, not days.

⏱ **Time Saved:** 1-3 hours per listing (compared to traditional staging coordination)

☞ **Expertise Required:** ●●○○○ (Beginner friendly)

💰 **Cost vs. Benefit:** $15-40 for full listing vs. $2,000-5,000 for physical staging

The Ethical Implementation Process

Step 1: Understand the Disclosure Requirements

According to NAR guidance:

✅ **Permitted:** Removing temporary items (garden hose, moving boxes, clutter)

✅ **Permitted:** Adding furniture to empty rooms (virtual staging)

✅ **Permitted:** Sky replacement, lighting enhancement, color correction

✖ **Prohibited:** Removing or altering permanent structural elements

✖ **Prohibited:** Hiding defects, damage, or material property conditions

Standard: Would a reasonable buyer feel misled when they visit in person?

Step 2: Select Photos for Enhancement

Choose strategically:

- **Primary rooms:** Living room, primary bedroom, kitchen (buyers decide here)
- **Empty spaces:** Rooms that look barren or awkwardly proportioned when vacant
- **Odd layouts:** Spaces where furniture placement clarifies use
- **Exteriors:** When weather or lighting was poor during photo shoot

Don't virtually stage:

- Small secondary bedrooms (buyers see through it)
- Spaces with existing furniture (looks fake, raises questions)
- Photos where virtual furniture would block key features

Step 3: Use Your Chosen Tool

Based on your selection from Chapter 4:

Virtual Staging AI workflow:

1. Upload high-resolution empty room photo
2. Select room type from dropdown
3. Choose design style matching target buyer taste
4. Generate 2-3 variations
5. Select best option
6. Download both original and staged versions

ReimagineHome workflow:

1. Upload photo
2. Select "Virtual Staging" or specific enhancement

3. Choose style from presets or custom description

4. Generate and review options

5. Download and label clearly

Step 4: Implement with Proper Disclosure

In MLS:

- Caption: "Photos digitally staged" or "Virtual staging used"
- Include both staged and unstaged versions when possible

On your website and marketing:

- Clear labels on all virtually staged images
- Original photos available on request
- Disclosure statement in property materials

Sample disclosure language:

"Some photos in this listing feature virtual staging to help visualize the property's potential. Virtual staging is a digital enhancement, not representative of the current property condition. Original, unstaged photos are available upon request."

⚠ **CRITICAL:** Transparency builds trust. Deception destroys it. Always err on the side of disclosure.

📊 **Life Margin Impact:**

Beyond time savings, virtual staging removes a major point of stress: coordinating physical staging logistics, managing staging company schedules, worrying about theft or damage. It's mental load reduction as much as time reclamation.

Generating Video Tour Scripts

The Reality: Property videos drive 400% more inquiries than listings without video. But scripting compelling narration is time-consuming and most agents wing it (resulting in rambling, unfocused tours).

The Opportunity: AI generates structured, professional scripts in 3 minutes. You refine for 5 minutes. You shoot with confidence.

⏱ **Time Saved:** 45-60 minutes per video

◎ **Expertise Required:** ●●○○○ (Beginner friendly)

The Professional Video Script Workflow

Step 1: Map Your Video Journey (5 minutes)

Before prompting, determine:

- **Video length target** (90 seconds? 2.5 minutes? 4 minutes?)
- **Logical flow through property** (exterior → main living → bedrooms → special features → exterior)
- **Key features to emphasize** (3-5 standout elements)
- **Opening hook** (what grabs attention in first 5 seconds?)
- **Closing call to action** (what do you want viewers to do?)

Step 2: Use the Tested Prompt

> Create a [LENGTH]-minute video tour script for a [PROPERTY TYPE] in [LOCATION].

The tour should flow in this order:

1. [SPACE/AREA] ([APPROXIMATE TIME IN SECONDS])

2. [SPACE/AREA] ([APPROXIMATE TIME IN SECONDS])

3. [SPACE/AREA] ([APPROXIMATE TIME IN SECONDS])

[Continue for all major spaces]

Key features to highlight:

- [FEATURE 1 with brief context]

- [FEATURE 2 with brief context]

- [FEATURE 3 with brief context]

[List 4-6 key features]

Target audience: [BUYER PROFILE]

Include:

- A compelling opening hook that addresses [SPECIFIC BUYER NEED]

- Natural transitions between spaces

- Lifestyle benefits, not just feature lists

- A clear call to action at the end

The tone should be [PROFESSIONAL/WARM/ENTHUSIASTIC/SOPHISTICATED] and match the property's character.

Include brief pauses marked with [PAUSE] where visual elements should speak for themselves.

Step 3: Refine the Script (5-7 minutes)

Review for:

- **Natural flow** (does it match how someone walks through?)
- **Timing** (read it aloud, adjust for your speaking pace)
- **Key feature emphasis** (did AI highlight what matters most?)
- **Authentic voice** (does this sound like you, or generic?)
- **Strategic positioning** (does it address target buyer concerns?)

Step 4: Practice and Record

- Read through 2-3 times before recording
- Record in segments (easier to edit, more natural delivery)
- Leave pauses for visual emphasis
- Re-record any section that feels forced

Sample Output (excerpt):

[EXTERIOR - 15 seconds]

> "Welcome to 847 Riverside Drive, where mid-century architecture meets modern desert luxury. [PAUSE] Built in 1962 and thoughtfully renovated in 2024, this Palm Springs gem captures everything you love about desert modernism."

[LIVING AREA - 25 seconds]

> "Step inside to walls of glass that blur the line between indoor and outdoor living. Original terrazzo floors flow throughout the main living

area, while ten-foot ceilings create an airy, expansive feel. [PAUSE] Notice how every room frames the San Jacinto mountains perfectly."

 Human Touch Required: Your local knowledge and on-site observations make scripts go from good to great. AI provides structure. You provide insight.

Section 2: Client Communications

Personalized Email Templates for Every Stage

The Reality: You send 30-50 emails per week. Many follow similar patterns: post-showing follow-ups, transaction updates, market insights, check-ins. Each takes 10-15 minutes to compose from scratch because you're trying to sound personal and professional simultaneously.

The Opportunity: AI generates email drafts in 60 seconds that you personalize in 2-3 minutes. Same quality, one-fifth the time.

 Time Saved: 8-12 minutes per email (3-5 hours weekly for active agents)

 Expertise Required: ●●○○○ (Beginner friendly)

 Human Touch Required: ●●●●○ (High, personalization is critical)

The Email Library System

Rather than prompting each email individually, build a library of templates for recurring scenarios. Then customize each use.

Step 1: Map Your Email Touchpoints

Buyer Journey Emails:

- Initial inquiry response
- Post-consultation follow-up
- Pre-showing property details

- Post-showing feedback request
- Offer preparation guidance
- Under contract milestone updates
- Pre-closing preparation

Seller Journey Emails:

- Initial consultation follow-up
- Pre-listing preparation checklist
- Listing launch announcement
- Weekly activity reports
- Showing feedback summary
- Offer discussion
- Under contract updates

Long-Term Relationship Emails:

- Quarterly market updates
- Home maintenance tips (seasonal)
- Anniversary of purchase/sale
- Referral requests
- Event invitations

Step 2: Generate Your Core Templates

Template Creation Prompt:

> Write an email template for a real estate agent to send to a [BUYER/SELLER] client who [SPECIFIC SITUATION].

The email should:

- Have a clear, specific subject line

- Open with a personalized greeting

- [SPECIFIC CONTENT REQUIREMENTS]

- Address common questions about [TOPIC]

- Include a clear call to action

- Close warmly and professionally

The tone should be [PROFESSIONAL/WARM/URGENT/SUPPORTIVE], approximately [WORD COUNT] words.

My communication style is [DESCRIBE YOUR VOICE, e.g., "knowledgeable but approachable, focused on being a trusted advisor who explains clearly without talking down"].

Include [BRACKETS] for fields I need to customize for each client.

Step 3: Build Your Template Library

Create 8-12 core templates covering your most frequent scenarios. Save them in:

- Your CRM's template section
- A Google Doc or Notion page
- Your email client's template feature

Use a clear naming system:

- BUY_01_Initial_Inquiry_Response
- BUY_02_Post_Showing_Followup
- SELL_01_Listing_Launch
- NURTURE_Quarterly_Market_Update

Real Template Example: Post-Showing Follow-Up

The Prompt:

> Write an email template for a real estate agent to send to a buyer client 24 hours after showing them 3-5 properties.

The email should:

- Have a clear, specific subject line

- Thank them for their time

- Ask for specific feedback on each property viewed

- Offer to answer any questions that have come up

- Suggest logical next steps based on their interest level

- Maintain momentum without being pushy

The tone should be professional yet conversational, approximately 200 words.

My communication style is knowledgeable but approachable, focused on being a trusted advisor who listens carefully and guides thoughtfully.

Include [BRACKETS] for fields I need to customize for each client.

The Generated Template:

Subject: Thoughts on yesterday's showings?

Hi [CLIENT_NAME],

Thank you for taking the time yesterday to view properties together. I enjoyed hearing your thoughts as we walked through each one.

I'd love to get your feedback on what we saw:

[PROPERTY_1_ADDRESS]: What stood out to you (positively or negatively)?

[PROPERTY_2_ADDRESS]: How did this one compare to your expectations?

[PROPERTY_3_ADDRESS]: Any features here that particularly appealed to you?

Your reactions help me understand what matters most to you and refine our search to find the right fit.

If any questions have come up since yesterday, I'm here to discuss. I'm also happy to schedule viewings for [ADDITIONAL_PROPERTIES] or revisit anything we saw.

What would be most helpful as a next step?

Best regards,

[YOUR_NAME]

[CONTACT_INFO]

Step 4: Customize Each Use (2-3 minutes)

Replace bracketed fields with:

- Client's actual name and personal details
- Specific property addresses and features discussed
- Your observations from the showing
- Any follow-up items you discussed
- Timeline considerations unique to this client

Add a personal touch:

- Reference something they mentioned ("I followed up on that school rating you asked about...")
- Connect to their stated priorities ("Given your interest in outdoor space, I also found...")
- Show you were listening ("I know you mentioned wanting a home office...")

⚠ **NEVER send AI-generated emails without personalization.** Generic templates destroy the relationship advantage you're building.

📊 **Life Margin Impact:**

If you send 40 emails weekly and save 10 minutes each, that's 400 minutes (6.7 hours) weekly. That's 348 hours annually. That's nearly 9 full work weeks of time reclaimed. What matters enough to you to deserve 9 extra weeks per year?

🔖 **Future-Ready Checkpoint:**

The email system you're building (trigger-based templates with personalization) is a solid first step in how you'll manage AI email agents to amplify your communications. Those systems will draft, personalize based on CRM data, and queue for your approval. Agents who understand template systems and personalization now will seamlessly upgrade to supervising autonomous communication agents.

Social Media Content Creation at Scale

The Reality: Consistent social presence builds credibility. But creating 3-5 quality posts weekly is exhausting. Most agents start strong in January, fade by March, ghost their accounts by summer.

The Opportunity: Generate two weeks of content in 30 minutes. Customize each post in 2-3 minutes. Schedule and forget. Consistency without burnout.

⏱ **Time Saved:** 2-3 hours per week

◉ **Expertise Required:** ●●○○○ (Beginner friendly)

The Batch Content Creation System

Step 1: Define Your Content Strategy

Before generating content, clarify:

Posting frequency: 3-5 posts weekly (sustainable for most agents)

Content categories (sample mix):

- 40% Educational/valuable (market insights, tips, process guidance)
- 30% Listings and success stories
- 20% Personal brand/behind-the-scenes
- 10% Community and engagement

Target platforms:

- Instagram (visual, lifestyle-focused)
- Facebook (community, detailed)
- LinkedIn (professional, market analysis)

Step 2: Generate a Content Calendar

Bi-Weekly Calendar Prompt:

> Create a 2-week social media content calendar with [NUMBER] posts per week for a real estate agent.
>
> My business focus: [YOUR NICHE, e.g., "first-time homebuyers in suburban Chicago" or "luxury waterfront properties in Charleston"]
>
> Content mix:
>
> - [NUMBER] educational posts (buyer/seller tips, market insights, process guidance)
>
> - [NUMBER] listing showcases or success stories
>
> - [NUMBER] personal brand or community involvement posts
>
> For each post, provide:
>
> - Platform recommendation (Instagram, Facebook, or LinkedIn)
>
> - Post copy (100-150 words)
>
> - 5-7 relevant hashtags
>
> - Visual content suggestion
>
> - Clear call to action
>
> The tone should be [CONVERSATIONAL/AUTHORITATIVE/HELPFUL] and position me as a knowledgeable local expert who genuinely cares about helping people make informed decisions.
>
> Focus on providing genuine value, not being promotional. Educational content should be actionable and specific.

Step 3: Customize and Personalize

AI gave you structure and ideas. Now make it yours:

- **Add local specificity** ("In Naperville, spring sellers typically list in late March...")
- **Include recent examples** ("Just closed on a property where the buyer...")
- **Update for current events** (interest rate changes, local market shifts)
- **Match your actual voice** (would you really say this?)

- **Add authentic photos** (your listings, your community, your face)

Step 4: Create Visual Content

Use Canva Pro's AI features:

- Magic Design: Upload a photo, get 10+ post layouts instantly
- Magic Write: Generate complementary captions
- Background Remover: Clean product shots for property features
- Brand Kit: Maintain consistent fonts, colors, logos

Time investment: 15-20 minutes for 10-12 graphics once you have the copy

Step 5: Schedule and Monitor

Use free or low-cost scheduling tools:

- Meta Business Suite (Instagram + Facebook, free)
- Later (free for basic scheduling)
- Buffer (free tier available)

Schedule in one sitting: Sunday evening or Monday morning, load two weeks of content

Daily engagement time: 15 minutes to respond to comments and DMs

Sample Post: Educational Content

Platform: Instagram

Topic: First-time buyer preparation

Generated Post:

> Thinking about buying your first home this year? Start with these three financial steps today:

1 Check your credit score (free at annualcreditreport.com). Scores above 740 typically qualify for the best rates. If yours is lower, spend 6 months improving it before shopping.

2 Calculate your real budget using the 28/36 rule: housing costs shouldn't exceed 28% of gross income, total debt shouldn't exceed 36%. Online calculators help, but talk to a lender for your specific scenario.

3 Save beyond the down payment. Budget for closing costs (2-5% of purchase price), moving expenses, immediate repairs, and 3-6 months of reserves.

Questions about the homebuying process in [YOUR AREA]?

Drop them below or DM me directly.

#firsttimehomebuyer   #homebuyingtips   #realestatetips
#[yourcity]realestate   #mortgagetips   #financialplanning
#homeownership

Visual suggestion: Clean graphic with the three steps as text overlay on a photo of happy homeowners with keys

Customization needed:

- Replace [YOUR AREA] with your specific market
- Add your local lender relationships or preferred resources
- Update if your market has specific first-time buyer programs
- Use a photo from one of your actual first-time buyer closings (with permission)

Just-in-Time Learning: "Batch vs. Real-Time Content"

Batch-created content maintains consistency. Real-time content captures opportunities. The ideal ratio is 80% scheduled batch content, 20% real-time posts reacting to market news, new listings, or community events. Batch creation gives you the foundation. Real-time additions keep you relevant.

Section 3: Property Research & Analysis

Creating Comparative Market Analyses (CMAs)

The Reality: A quality CMA takes 90-120 minutes: pulling comps, analyzing adjustments, formatting for clients, writing narrative insights. You do 2-4 per month. It's essential but time-intensive.

The Opportunity: AI analyzes comps and identifies patterns in 5 minutes. You provide professional interpretation and strategic recommendations in 15-20 minutes. Total time: 25-30 minutes for better quality.

⏱ **Time Saved:** 60-90 minutes per CMA

◎ **Expertise Required:** ●●●●○ (Advanced, requires professional judgment)

🖐 **Human Touch Required:** ●●●●● (Very high, this is core professional expertise)

The Professional CMA Workflow

Step 1: Gather Comparable Data (15 minutes)

Pull from your MLS:

- 6-8 sold comps (closed within 90-120 days)
- 3-4 active listings (current competition)
- 2-3 pending sales if available (market direction indicator)

For each comp, note:

- Address, sale price, sale date
- Beds, baths, square footage, lot size
- Year built, condition, updates
- Key features, garage, special amenities
- Days on market, list-to-sale ratio

Pro tip: AI can help gather some of this data from public sources like Zillow, Redfin, or Realtor.com by analyzing screenshots or data you paste. However, always verify against MLS data before using. Public sites can have outdated or inaccurate information.

Step 2: AI-Assisted Analysis

Analysis Prompt:

> Help me analyze these comparable properties for a [PROPERTY TYPE] in [NEIGHBORHOOD].

Subject Property:

- [BEDS] beds, [BATHS] baths

- [SQUARE FOOTAGE] square feet

- Built in [YEAR]

- Key features: [LIST NOTABLE FEATURES]

- Condition: [EXCELLENT/GOOD/AVERAGE/NEEDS WORK]

Comparable Properties:

1. [ADDRESS], sold [PRICE] on [DATE]

  - [BEDS] beds, [BATHS] baths, [SQFT] sqft

  - Features: [KEY FEATURES]

  - Condition: [CONDITION]

  - Days on market: [DOM]

2. [Repeat for 5-8 comps]

Based on these comparables:

1. What is the indicated price range for my subject property?

2. What adjustments should be made for key differences (size, condition, features, location)?

3. Which comparable provides the best basis for valuation and why?

4. What market trends are evident from this data (days on market, list-to-sale ratios, price direction)?

5. What additional information would strengthen this analysis?

Provide your analysis in clear sections with specific reasoning for each conclusion.

Step 3: Professional Review and Enhancement (15-20 minutes)

AI identified patterns and performed calculations. Now apply your expertise:

- **Verify the math** (AI can make calculation errors)
- **Apply local market knowledge** (school district boundaries, traffic patterns, upcoming development)
- **Consider factors AI can't quantify** (street appeal, neighborhood reputation, micro-location advantages)
- **Add strategic context** (is this a competitive pricing situation? Is there a timing consideration?)
- **Formulate your recommendation** (not just a number, but a strategic price position)

Step 4: Create Client-Friendly Presentation

Presentation Prompt:

> Create a client-friendly summary of this comparative market analysis for a [BUYER/SELLER].

Key findings from my analysis:

- Recommended price range: [RANGE]

- Most relevant comparables: [LIST 3 TOP COMPS WITH BRIEF REASON]

- Key market trends: [TRENDS IDENTIFIED]

- Strategic considerations: [YOUR PROFESSIONAL INSIGHTS]

The presentation should:

- Open with the recommended price and clear rationale

- Explain methodology in simple, non-technical terms

- Highlight the most relevant comparables with brief explanations

- Address potential objections or questions

- Include a market conditions summary

- Close with next steps

Use clear language appropriate for a client who isn't a real estate professional. Avoid jargon. Build confidence in the analysis.

Length: approximately 400-500 words.

Step 5: Add Visual Enhancements

Create supporting visuals using:

- Comparison chart (subject vs. best 3-4 comps)
- Price trend graph (last 6-12 months in neighborhood)
- Days-on-market comparison
- Map showing comp locations relative to subject

Canva Pro makes this fast: 10-15 minutes for all graphics using templates

! PROFESSIONAL RESPONSIBILITY CHECKPOINT:

You remain responsible for valuation accuracy and strategic pricing recommendations. AI processes data and identifies patterns. You provide professional judgment, market expertise, and strategic guidance. Never present AI analysis as your definitive recommendation without thorough professional review.

ᴸᴵ Life Margin Impact:

Saving 70 minutes per CMA × 3 CMAs monthly = 210 minutes (3.5 hours) monthly, or 42 hours annually. But there's a hidden benefit: when CMAs take 25 minutes instead of 2 hours, you're more willing to do them. That means more listing presentations, which means more business. The time savings unlock opportunity, not just reclaim hours.

Client-Friendly Market Reports

The Reality: Clients ask "What's the market doing?" You know the data (you track it), but formatting it into clear, professional reports is tedious. Most agents send occasional updates rather than consistent, valuable market intelligence.

The Opportunity: Generate monthly market reports in 20-25 minutes that position you as the local market authority. Consistent value builds trust and top-of-mind awareness.

⏱ **Time Saved:** 1.5-2 hours per report

☞ **Expertise Required:** ●●●○○ (Intermediate)

🖐 **Human Touch Required:** ●●●●○ (High, interpretation is key value)

The Market Report System

Step 1: Gather Current Data (10 minutes)

Collect from your MLS or association:

- Active listing count (compare to last month, last year)
- Median sales price (monthly, year-over-year)
- Average days on market (trend direction)
- Months of inventory (buyer's/seller's/balanced market indicator)
- List-to-sale price ratio (pricing accuracy indicator)
- Interest rate environment (current, recent trend)

Step 2: Generate Initial Analysis

Market Analysis Prompt:

> Help me interpret this real estate market data for [NEIGHBORHOOD/CITY] to create a client-friendly market report for [MONTH/YEAR].

Current Data:

- Active listings: [NUMBER] (up/down [X]% from last year)

- Median sales price: $[PRICE] (up/down [X]% from last year)

- Average days on market: [NUMBER] (up/down [X]% from last year)

- Months of inventory: [NUMBER]

- List-to-sale price ratio: [X]%

- 30-year mortgage rate: [X]% (compared to [X]% last quarter)

Additional Context:

- [LOCAL ECONOMIC FACTORS, if any]

- [SEASONAL CONSIDERATIONS]

- [NOTABLE TRENDS YOU'VE OBSERVED]

Create an analysis that:

1. Identifies current market condition (buyer's market, seller's market, or balanced)

2. Explains the most significant trends in simple language

3. Provides specific implications for buyers and for sellers

4. Offers 2-3 strategic recommendations for each group

5. Predicts likely direction over next 3-6 months (with appropriate caveats about uncertainty)

Write in clear, confident language that helps clients make informed decisions without creating alarm or hype. Approximately 300-400 words.

Step 3: Enhance with Your Expertise (10 minutes)

Add what AI can't provide:

- **Hyperlocal nuance** (micro-market differences, neighborhood-specific trends)
- **Anecdotal evidence** (what you're seeing in showings, buyer behavior shifts)
- **Strategic context** (how this fits into larger economic picture)
- **Forward-looking insight** (upcoming developments, seasonal patterns, predicted shifts)

Step 4: Create Visual Data Story (15 minutes)

Data is more digestible with good visuals:

Chart ideas:

- Median price trend (12-month line graph)
- Inventory levels (bar chart comparing year-over-year)
- Days on market (trend line)
- Sales volume (monthly comparison)

Canva Pro workflow:

1. Use "Chart" elements
2. Input your data
3. Apply your brand colors
4. Export as image
5. Insert into report

Step 5: Format for Distribution

Format options:

- **PDF Report:** Professional, saveable, sharable (ideal for email)
- **Social Media Graphics:** Key stats as Instagram carousel or Facebook post
- **Video Script:** Record 90-second video walking through highlights
- **Email Newsletter:** Text + embedded graphics

Distribution channels:

- Email database (monthly)
- Social media (quarterly or when significant shifts occur)
- Website blog (SEO benefit + evergreen value)
- Print mailers to farm area (for geographic specialists)

Sample Report Excerpt:

> Naperville Real Estate Market Update - December 2025

Market Snapshot: Our local market continued its shift toward balanced conditions this month. With 4.2 months of inventory (up from 3.1 last December), we're seeing more choices for buyers and more realistic pricing from sellers.

Key Trends:

Median home price: $585,000 (up 3.2% year-over-year, but down 1.8% from our October peak)

Average days on market: 32 days (compared to 21 days last December)

Homes are selling for 98.1% of list price (down from 99.4% last year)

What This Means:

For Buyers: You have more negotiating power than you did 6-12 months ago. Homes are sitting longer, which means sellers are more motivated. However, well-priced properties in desirable areas still receive multiple offers. Strategic pricing knowledge is crucial.

For Sellers: Pricing accuracy matters more than ever. Overpriced homes sit while fairly priced properties still sell quickly. Professional staging and marketing make the difference between 30 days and 60+ days on market.

Looking Ahead: Expect the balanced market to continue through spring. Interest rates have stabilized around 6.5%, which brings more buyers off the sidelines. Spring inventory will be critical to watch.

Just-in-Time Learning: "Data vs. Insight"

Data tells what happened. Insight explains why it matters and what to do about it. AI provides excellent data synthesis. Your value is interpreting that data for your specific clients' situations. Never send raw data without interpretation.

Section 4: Transaction Management

Document Review & Summarization

The Reality: Contracts, disclosures, inspection reports, and appraisals contain critical information buried in dense text. You need to extract key points, identify concerns, and explain implications to clients. Each document takes 30-45 minutes to review thoroughly and summarize.

The Opportunity: AI can identify key provisions, flag unusual terms, and create initial summaries in 3-5 minutes. You spend 10-15 minutes on focused professional review. Total: 15-20 minutes for more thorough analysis.

⏱ **Time Saved:** 20-30 minutes per document

◎ **Expertise Required:** ●●●●○ (Advanced, requires professional judgment)

👆 **Human Touch Required:** ●●●●● (Very high, professional review essential)

The Professional Document Review Workflow

Step 1: Upload and Scan

Most modern AI tools (ChatGPT, Claude, Gemini) accept PDF uploads. Upload your document and use focused prompts.

Initial Review Prompt:

> Review this [DOCUMENT TYPE: purchase agreement/inspection report/disclosure statement] for a real estate transaction in [STATE].
>
> As you review:
>
> 1. Identify and list all key terms, conditions, and dates
>
> 2. Flag any unusual or potentially problematic clauses
>
> 3. Note all deadlines and contingency dates in chronological order
>
> 4. Highlight provisions that differ from standard [STATE] contracts
>
> 5. Identify areas where additional negotiation may be beneficial
>
> Create a structured summary with:
>
> - Critical dates and deadlines (in chronological order)
>
> - Key financial terms
>
> - Contingencies and conditions
>
> - Unusual provisions or concerns
>
> - Action items requiring immediate attention

This summary will guide my focused professional review, not replace my obligation to read the entire document carefully.

Step 2: Focused Professional Review (10-15 minutes)

Use AI summary as your roadmap, but conduct your own thorough review:

- **Read the entire document** (AI might miss context-dependent clauses)
- **Verify all dates and figures** AI identified
- **Apply your professional judgment** to flagged concerns
- **Consider jurisdiction-specific requirements** (AI may not know local practice)
- **Note anything AI missed** (unusual contingencies, handwritten additions)
- **Assess strategic implications** (does this favor one party?)

Step 3: Create Client Explanation

Client Summary Prompt:

> Create a client-friendly explanation of this [DOCUMENT TYPE] for my [BUYER/SELLER] client.

Key provisions I've identified:

- [LIST MOST IMPORTANT TERMS]

- [NOTABLE CONTINGENCIES]

- [CRITICAL DATES]

- [AREAS OF CONCERN OR OPPORTUNITY]

The explanation should:

- Use plain language without legal jargon

- Highlight the most important obligations and rights

- Explain any unusual or negotiated terms

> - List all upcoming deadlines and required actions in chronological order
>
> - Address likely client questions
>
> - Include appropriate disclaimers about seeking legal/professional advice where needed
>
> The tone should be clear, educational, and confidence-building while being honest about risks or concerns.
>
> Approximately 300-400 words.

Step 4: Schedule Follow-Up Discussion

Send the written summary, then schedule a call to:

- Answer questions
- Provide strategic counsel
- Confirm understanding of obligations
- Discuss any concerns flagged

Never rely on written explanation alone for complex documents.

⚠ PROFESSIONAL RESPONSIBILITY CHECKPOINT:

This is perhaps the most critical checkpoint in the entire book. NEVER rely solely on AI for legal document review. AI assists with organization and initial analysis. YOU are responsible for accurate document review, disclosure verification, and contract compliance. This 10-15 minute human verification protects your career, your license, and your clients' interests. There is no shortcut here.

The Story You Don't Want Becoming Real:

You used AI to review a purchase agreement. AI correctly identified most terms but missed a handwritten addendum changing the closing date by two weeks. You didn't read the full document. The mistake caused a cascade of problems: seller's relocation timeline disrupted, buyer's rate lock expiration, loan delays. Your E&O carrier ended up paying $18,000 to resolve the dispute.

The lesson: AI is a powerful assistant, not a replacement for professional review.

Transaction Timeline & Checklist Creation

The Reality: Managing 20+ deadlines and action items per transaction requires meticulous organization. Creating customized checklists for each transaction takes 30-40 minutes. Missing a deadline can tank a deal or expose you to liability.

The Opportunity: Generate comprehensive, customized timelines in 5 minutes. Review and adjust in 5 minutes. Total: 10 minutes for better organization.

⏱ **Time Saved:** 20-30 minutes per transaction

◉ **Expertise Required:** ●●●○○ (Intermediate)

🌀 **Human Touch Required:** ●●●○○ (Medium, verification needed)

The Timeline Creation System

Step 1: Extract Key Dates

From your executed contract, identify:

- Effective date
- Earnest money deadline
- Inspection period end
- Financing contingency deadline
- Appraisal deadline
- Title commitment deadline
- Final walkthrough date
- Closing date
- Possession date

Step 2: Generate Comprehensive Timeline

Timeline Prompt:

> Create a comprehensive transaction timeline for a [RESIDENTIAL PURCHASE/SALE] in [STATE].
>
> **Key Contract Dates:**
>
> - Effective date: [DATE]
>
> - Earnest money due: [DATE]
>
> - Inspection period ends: [DATE]
>
> - Financing approval deadline: [DATE]
>
> - Appraisal deadline: [DATE]
>
> - Closing date: [DATE]
>
> **Additional Requirements:**
>
> - [ANY SPECIAL STIPULATIONS]
>
> - [LOCAL CLOSING PRACTICES]
>
> - [UNIQUE TRANSACTION CONSIDERATIONS]
>
> Create a timeline that:
>
> 1. Lists all events chronologically with specific dates
>
> 2. Identifies responsible party for each action (buyer, seller, agent, lender, attorney, etc.)
>
> 3. Distinguishes business days vs. calendar days where relevant
>
> 4. Includes standard milestones not explicitly in contract (title ordered, survey completed, utilities scheduled, etc.)
>
> 5. Notes critical path dependencies (what must happen before what)
>
> 6. Flags high-priority deadlines where missing them has serious consequences
>
> Format as a detailed calendar with weekly sections and milestone markers clearly visible.

Step 3: Generate Role-Specific Checklists

Checklist Prompt:

> Create a detailed action checklist for the [BUYER/SELLER/AGENT] based on this transaction timeline.
>
> [PASTE THE TIMELINE YOU JUST GENERATED]
>
> The checklist should:
>
> 1. Be organized chronologically by week or phase
>
> 2. Include every action the [ROLE] needs to take
>
> 3. Note all documents to provide and deadlines to meet
>
> 4. Include preparation steps before major milestones
>
> 5. Flag items that are easily overlooked
>
> 6. Indicate which items are time-sensitive vs. general preparation
>
> Format as an actionable checklist with checkboxes and clear deadlines.

Step 4: Integrate with Your Systems

Transfer to:

- Transaction management software (Dotloop, Skyslope, etc.)
- Digital calendar with automated reminders
- CRM task list
- Client portal (if you use one)

Set up reminders:

- 7 days before each major deadline
- 3 days before each major deadline
- Day before each major deadline

Share with client:

- Email PDF of their specific checklist
- Grant access to shared calendar
- Weekly check-in emails with upcoming items

Step 5: Monitor and Update

Transactions rarely go exactly as planned. Update your timeline when:

- Deadlines are extended
- New contingencies are added
- Closing date changes
- Issues arise requiring attention

Life Margin Impact:

The life margin benefit here isn't just time saved. It's stress eliminated. When you have comprehensive timelines and reminders, you stop worrying about what you might have forgotten. You sleep better. You're more present with clients because you're confident nothing is slipping through the cracks. That peace of mind is worth more than the 25 minutes saved per transaction.

Section 5: Business Development

Lead Qualification & Response Systems

The Reality: Every lead needs a prompt response, but not every lead deserves equal time. You spend 15-20 minutes per lead on initial outreach, qualification questions, and follow-up scheduling. Many leads aren't ready, qualified, or serious. Time investment often exceeds return.

The Opportunity: Create systematic qualification that takes 5-7 minutes per lead while being more thorough and consistent. Identify your A-leads quickly and nurture B/C leads automatically.

⏱ **Time Saved:** 10-15 minutes per lead (5-7 hours weekly for agents generating 30-40 leads monthly)

☺ **Expertise Required:** ●●●○○ (Intermediate)

🖐 **Human Touch Required:** ●●●●○ (High for qualified leads, medium for nurture sequences)

The Lead Management System

Step 1: Define Your Qualification Criteria

Create clear criteria for lead prioritization:

A-Leads (immediate personal follow-up):

- Buying/selling within 60-90 days
- Pre-approved or financially ready
- Specific property requirements
- Referral from past client or sphere
- High engagement (multiple property views, form submissions)

B-Leads (systematic nurture sequence):

- Buying/selling within 3-9 months
- Needs financial preparation
- General interest, not specific properties yet
- Good communication response

C-Leads (long-term educational content):

- Exploring/researching (12+ months out)
- Not yet financially qualified
- Limited communication response
- General market curiosity

Step 2: Create Response Templates

Initial Response Prompt (A-Lead):

> Generate a warm, professional initial response email for a new lead who [SPECIFIC ACTION: requested property information/submitted contact form/called about listing].

The email should:

- Acknowledge their specific inquiry immediately

- Provide [INITIAL VALUE: property details/requested information]

- Ask 2-3 conversational qualifying questions about timeline, readiness, and specific needs

- Communicate your value proposition clearly

- Offer multiple easy ways to continue the conversation (call, text, email, in-person)

- Create gentle urgency without being pushy

The tone should be enthusiastic and helpful, approximately 150-200 words.

Include [BRACKETS] for information to customize per lead.

Step 3: Develop Scoring System

Scoring Prompt:

> Help me create a simple 1-5 lead scoring system for real estate prospects.

Factors to consider:

- Timeframe to transaction

- Financial readiness

- Decision-making clarity (knows what they want vs. just browsing)

- Communication responsiveness

- Specific needs vs. general browsing

- Referral source quality

For each factor, define scoring criteria (what = 1 point, what = 5 points).

Then create a simple formula to calculate an overall lead score from these factors.

Finally, suggest action protocols for each score range:

- Score 20-25: [immediate action]

- Score 15-19: [action]

- Score 10-14: [action]

- Score 5-9: [action]

Step 4: Build Automated Nurture Sequences

Nurture Sequence Prompt:

> Create a 60-day nurture email sequence for a [BUYER/SELLER] lead who is planning to [SPECIFIC GOAL] in approximately [TIMEFRAME].

The sequence should include:

- Email frequency (timing for each message: Day 1, Day 4, Day 7, etc.)

- Subject line for each email

- Brief content summary for each message

- Value provided in each contact (no pure check-ins without value)

- Varied formats (market updates, tips, success stories, resources)

- Clear opportunities for lead to re-engage when ready

- Natural off-ramps if lead is not responsive

Assume a gradual nurturing approach that demonstrates expertise and builds trust without being pushy or salesy.

Generate detailed outlines for 8-10 emails spread over 60 days.

Step 5: Implement in Your CRM

Most modern CRMs support automation:

- Set triggers based on lead source and behavior
- Assign leads to sequences based on scores
- Queue responses for your review before sending
- Track engagement (opens, clicks, responses)
- Adjust sequences based on behavior

⚠ **Never set fully automated responses without review for high-value leads.** A-leads deserve personal, customized attention.

🌰 **Human Touch Required:**

For A-leads: Full personalization, immediate personal response, custom recommendations.

For B-leads: Light personalization (name, specific interest), review automated messages before sending.

For C-leads: Automated educational content is appropriate, with quarterly personal check-ins.

📝 **Future-Ready Checkpoint:**

The lead management system you're building (qualification criteria, scoring, segmented nurture tracks) sets the foundation for agentic AI lead management in just a few years. Those systems will score leads, research prospects, draft personalized outreach, and schedule appointments automatically. Agents who understand lead qualification and nurture strategy now will seamlessly supervise autonomous business development agents.

Productivity & Time Optimization

The Reality: Most agents struggle with time management not because they're lazy or inefficient, but because real estate demands both reactive (client emergencies, showing requests, offer negotiations) and proactive work (prospecting, content creation, business development). Reactive work always wins. Proactive work dies.

The Opportunity: Use AI to analyze your time allocation, optimize your schedule, and create time-blocking systems that protect your most valuable activities.

⏱ **Time Reclaimed:** 5-10 hours per week through better allocation and protection of high-value time

☺ **Expertise Required:** ●●○○○ (Beginner friendly)

Human Touch Required: ●●●○○ (Medium, requires personal discipline)

The Time Optimization System

Step 1: Audit Current Reality (one week)

Track everything for 7 days:

- Client calls/meetings (scheduled and reactive)
- Property showings
- Transaction coordination
- Marketing creation
- Lead generation/prospecting
- Administrative tasks
- Email/communication management
- "Wasted" time (unproductive browsing, inefficient workflows)

Use simple tracking:

- Timer app (Toggl, RescueTime)
- Paper log (just note start/end times)
- Calendar after-the-fact color coding

Step 2: Analyze with AI

Time Analysis Prompt:

> Help me analyze my time allocation for the past week as a real estate agent.

My current activities:

- Lead generation: [X] hours

- Client meetings/calls: [X] hours

- Property showings: [X] hours

- Transaction coordination: [X] hours

- Marketing creation: [X] hours

- Administrative tasks: [X] hours

- Email/communication: [X] hours

My business goals:

- [GOAL 1: e.g., increase transactions from 18 to 24 per year]

- [GOAL 2: e.g., build stronger referral pipeline]

- [GOAL 3: e.g., improve work-life balance and reduce weekend work]

Constraints:

- Peak energy/focus time: [TIME OF DAY]

- Non-negotiable commitments: [SPECIFIC TIMES/DAYS]

- Preferred working hours: [TOTAL HOURS/WEEK]

Analyze my current allocation and suggest:

1. Where I'm under-investing given my goals

2. Where I'm over-investing without proportional return

3. Which activities should be batched for efficiency

4. What could be delegated, automated, or eliminated

5. How to protect my peak performance time for highest-value activities

Be specific and realistic. Don't suggest doubling my workweek. Show me how to reallocate for better results.

Step 3: Design Your Ideal Week

Time-Blocking Prompt:

> Create an optimized weekly time-blocking schedule for a real estate agent based on this analysis.

[PASTE YOUR TIME ANALYSIS AND GOALS]

The schedule should:

1. Allocate specific blocks for each major activity type

2. Protect peak performance time for high-cognitive work (CMAs, strategic planning, content creation)

3. Batch similar activities (all showings on certain days, admin blocks, etc.)

4. Include buffer time for unexpected issues (at least 20% of schedule)

5. Balance reactive client needs with proactive business building

6. Maintain boundaries for personal/family time

7. Include weekly planning and review time

Create a week-at-a-glance schedule with specific time blocks for Monday-Friday. Format as a table for easy reference.

Step 4: Implement Gradually

Don't try to perfectly follow your ideal schedule immediately. That's a recipe for frustration and abandonment.

Week 1: Implement one major time block (e.g., "Tuesday and Thursday mornings are for showings only")

Week 2: Add another block (e.g., "Monday 9-11am is for content creation, no client calls")

Week 3: Add boundaries (e.g., "No weekend showings except by appointment made by Thursday")

Week 4: Fine-tune based on reality

Step 5: Protect Your Time with Systems

Create friction for interruptions:

- Calendar booking links (Calendly, etc.) that show only your available time blocks
- Auto-responders during focus blocks ("In focused work session, will respond by [TIME]")
- Clear "yes/no" criteria for exceptions ("I take same-day appointments only for pre-approved buyers with offers")

Build supporting habits:

- Sunday evening: plan the week ahead
- Daily morning: review today's priorities
- Daily evening: document time spent (5 min)
- Friday afternoon: weekly review (what worked, what didn't)

Sample Weekday Schedule (Foundation Agent):

Time	Mon.	Tue.	Wed.	Thu.	Fri.
9-11	Marketing /Content	Show-ings	Admin/ Email	Show-ings	Planning/ CMA
11-1	Client Calls	Show-ings	Listing Prep	Show-ings	Strategic Work
1-2	Lunch	Lunch	Lunch	Lunch	Lunch
2-4	Lead Follow-up	Trans-action Coord	Open House Prep	Trans-action Coord	Flex Time
4-6	Admin/ Email	Client Meetings	Client Meetings	Client Meetings	Weekly Review

Buffer time is built into "Flex Time" and scattered 15-min gaps.

Just-in-Time Learning: "Energy Management > Time Management"

Protecting time blocks matters, but protecting energy matters more. Schedule your highest-cognitive work (CMAs, strategic planning, complex negotiations) during your peak energy hours. Schedule routine work (email, admin, follow-ups) during lower-energy periods. Same amount of work, dramatically better output.

ıılLife Margin Impact:

Time blocking isn't just about efficiency. It's about reclaiming agency over your schedule. When you decide when showings happen instead of accepting every time a client proposes, you own your calendar. When you batch activities, you reduce context-switching mental fatigue. The result isn't just saved time, it's reduced stress and increased control. That's life margin: the space to live, not just work.

Putting It All Together: Your 30-Day Implementation Plan

You now have detailed workflows for every major function in your real estate practice. But knowing and doing are different.

Here's your roadmap to implement what you've learned:

Week 1: Choose and Master One Workflow

Monday: Review this chapter and select your highest-value workflow (biggest time drain or greatest frustration)

Tuesday-Thursday: Use that workflow 2-3 times, refining your process

Friday: Document your time savings and quality assessment

Target: Master one workflow completely

Week 2: Add a Second Workflow

Monday: Select your second-highest-value workflow

Tuesday-Thursday: Practice both workflows (Week 1 + Week 2)

Friday: Review progress, adjust prompts based on results

Target: Two workflows running smoothly

Week 3: Expand and Refine

Monday: Add a third workflow OR deepen one of your first two

Tuesday-Thursday: Focus on speed and quality improvements

Friday: Calculate total time saved across all workflows

Target: Sustainable 3-workflow system

Week 4: Scale and Systematize

Monday: Document your finalized prompts and processes

Tuesday-Wednesday: Train team members (if applicable) or create your personal playbook

Thursday: Identify next quarter's workflow additions

Friday: Celebrate wins, reflect on transformation

Target: Established system ready to expand

What You've Accomplished

By working through this chapter, you've:

✔ **Learned detailed workflows** for 12+ major real estate tasks

✔ **Received tested prompts** saving 20+ hours of trial and error

✔ **Understood where human judgment is essential** vs. where AI excels

✔ **Calculated potential time savings** of 8-15 hours weekly

✔ **Seen real examples** of successful AI-enhanced workflows

✔ **Built the foundation** for advanced automation (Chapter 6)

More importantly, you've transformed from someone who knows AI exists to someone who knows exactly how to deploy it across your entire practice.

You're no longer learning about AI in real estate. You're doing it.

Transition to Chapter 6

You've mastered individual workflows. Each one saves time and improves quality.

But the real power emerges when you connect these workflows into integrated systems. When a new lead triggers an automated qualification sequence that flows into personalized nurture emails that connect with transaction timelines that roll into post-closing relationship building—all with your strategic oversight at key decision points.

That's what Chapter 6 reveals: advanced automation workflows that create seamless client experiences while multiplying your capacity to serve more people excellently.

You've learned the instruments. Now we'll teach you to conduct the orchestra.

Chapter 6: Advanced Applications & Workflows

The System Sarah Built

Sarah had been using AI tools for six months. ChatGPT for descriptions. Virtual staging for photos. Email templates for follow-ups. Each tool saved time. But each tool also lived in its own silo.

Every Monday morning, she'd manually check her CRM for new leads. Copy their information into a spreadsheet. Research them on LinkedIn. Check county records for property ownership. Draft personalized outreach emails. Schedule follow-ups. Update her CRM. The process took 90 minutes.

Then she attended a workshop where another agent described something called "workflow automation." Not just using AI tools but connecting them into systems that worked together without constant human intervention.

Sarah spent two weeks learning. Another week building. The result: a connected system where new leads triggered automatic enrichment (pulling public property records, LinkedIn profiles, recent transaction history), scored themselves based on predefined criteria, generated personalized outreach emails referencing their specific situation, and queued high priority leads for her immediate attention with a complete research dossier.

Her Monday morning ritual dropped from 90 minutes to 12 minutes of reviewing what the system prepared.

But here's what surprised her most: the quality improved dramatically. The system never forgot to check a data source. It never sent generic emails because she was rushed. It never let a qualified lead sit uncontacted for three days because she was busy with closings.

That's when Sarah understood the difference between using AI tools and building AI-powered systems. Tools help you work faster. Systems multiply your capacity.

This chapter shows you how to build those systems.

The Shift from Tools to Systems

You've mastered individual workflows (Chapter 5). You understand how AI assists with specific tasks. Now we're elevating your thinking from "What can AI do for me?" to "How can connected AI systems run my business operations while I focus on the irreplaceable human work?"

This isn't about becoming a programmer or technical expert. It's about understanding how to identify, design, and supervise integrated workflows that transform your practice from manual orchestra to automated symphony.

Three levels of AI maturity:

Level 1: Task Assistance (Chapters 1-5)

Using AI to help with individual tasks. You prompt, AI responds, you use the output. Each task is separate. You're the operator.

Level 2: Workflow Automation (This chapter)

Connecting multiple AI-assisted tasks into sequences triggered by events. You design the system, AI executes the workflow, you supervise at checkpoints. You're the architect.

Level 3: Agentic AI (Chapter 9)

Autonomous systems that handle entire business functions with minimal supervision. You set goals and constraints, AI systems manage the details, you review outcomes. You're the strategist.

This chapter focuses on Level 2, while building the foundation for Level 3.

Future-Ready Checkpoint: Why Workflow Architecture Matters Now

The workflow design discipline you're learning now (identifying triggers, defining decision logic, setting quality checkpoints) will be required learning to manage semi-autonomous, agentic AI in as the technology continues to rapidly advance and mature. Agents who understand workflow architecture now will seamlessly transition to working alongside these systems. Those who don't will be overwhelmed by capabilities they don't know how to direct.

Understanding Workflow Components

Before building systems, understand the building blocks. Every automated workflow contains five core components:

1. Trigger

The event that starts the workflow automatically.

Examples:

- New lead submits contact form
- Contract reaches specific date milestone
- Property listing goes live
- Client emails with specific keyword
- Market data shows significant change

Your role: Define which events should trigger which workflows.

2. Data Source

Where the system pulls information to inform actions.

Examples:

- Your CRM database
- MLS listings and sales data

- Public property records
- County permit databases
- LinkedIn profiles
- Google search results
- Email conversation history

Your role: Identify which data sources provide valuable context for each workflow type.

3. Logic/Decision Points

Rules that determine what happens based on conditions.

Examples:

- IF lead score > 8, THEN notify agent immediately
- IF days on market > 30, THEN adjust marketing strategy
- IF client hasn't responded within 7 days, THEN send follow-up sequence
- IF property has a recent permit issued, THEN prioritize for investor outreach

Your role: Define the business rules that govern how the system behaves in different scenarios.

4. Actions

What the system does.

Examples:

- Send email with personalized content
- Update CRM fields
- Generate document or report
- Schedule appointment
- Post to social media
- Alert human agent for intervention

Your role: Specify what should happen at each stage, balancing automation with necessary human oversight.

5. Quality Checkpoints

Points where human review or approval is required.

Examples:

- Review generated client communication before sending
- Approve pricing recommendations
- Verify legal document accuracy
- Confirm unusual data patterns
- Authorize contract terms

Your role: Determine where automation is appropriate vs. where professional judgment is non-negotiable.

Just-in-Time Learning: "Automation Doesn't Mean Abdication"

The goal isn't to remove yourself from workflows entirely. It's to remove yourself from repetitive, low-value steps so you can focus your attention on high-value decision points. Well-designed automation systems increase the quality of your professional judgment by presenting you with better-prepared information at the right moments.

The Foundation: Client Journey Automation

The highest-ROI application of workflow automation is creating seamless client experiences across the entire transaction lifecycle. Let's break this down by journey stage.

Stage 1: Lead Acquisition & Qualification

The Manual Approach:

New lead comes in. You receive notification. You open their form submission. You google them. You check LinkedIn. You look up their property on the county website. You draft an email. You add them to your CRM. You set a reminder to follow up. Total time: 15-20 minutes per lead. Many leads slip through during busy periods.

The Automated System:

Trigger: Lead submits website form or inquiry

Data Sources Accessed:

- Form submission data
- Public property records (ownership, purchase date, mortgage info)
- LinkedIn profile (if available)
- Recent MLS activity in their area
- Your CRM historical data (have they inquired before?)

Logic Applied:

Calculate Lead Score (1-10):

+3 points: Owns property in target area

+2 points: Property owned 5+ years (equity built)

+2 points: Recent MLS browsing activity

+1 point: Complete contact information provided

+1 point: Specific timeline mentioned

+1 point: Referral source identified

IF Score >= 8: Route to "Hot Lead" workflow

IF Score 5-7: Route to "Warm Lead" workflow

IF Score < 5: Route to "Long-term Nurture" workflow

Actions Taken:

For **Hot Leads (8-10)**:

1. Immediate text/email to agent: "High-priority lead: [Name]"
2. Generate research dossier with all enrichment data
3. Draft personalized email referencing their specific property/situation
4. Queue for agent review and send within 15 minutes
5. Schedule a follow-up call for same day

6. Add to CRM with "Priority" tag

For **Warm Leads (5-7)**:

1. Send automated welcome email with valuable resource

2. Add to educational email sequence (buyer or seller track)

3. Notify agent via daily digest

4. Schedule follow-up for 48 hours

5. Add to CRM with timeline-based tagging

For **Long-term Nurture (1-4)**:

1. Send welcome email with general market information

2. Add to monthly newsletter list

3. Tag in CRM for quarterly personal check-in

4. Monitor for engagement increases (email opens, website returns)

Quality Checkpoint:

Agent reviews and approves personalized emails for Hot Leads before sending. Warm and Long-term communications use pre-approved templates with minimal customization needed.

ıl Life Margin Impact:

Before automation: 20 minutes per lead × 40 leads/month = 13.3 hours monthly

After automation: 3 minutes per lead review × 10 hot leads + 5 minutes daily digest review = 2.7 hours monthly

Time reclaimed: 10.6 hours monthly (127 hours annually)

But the larger impact is qualitative: zero leads fall through cracks during busy weeks, response time drops from hours to minutes, every lead receives appropriate attention level.

Stage 2: Active Buyer Journey

The Manual Approach:

After each showing, you text the client asking for feedback. You wait for their response. You log it in your notes. You search MLS for similar properties. You email them new options. You schedule the next showing. Repeat for every property.

The Automated System:

Trigger: Showing appointment marked complete in calendar

Data Sources Accessed:

- Showing feedback form (sent automatically post-showing)
- Client's property preference history
- MLS current listings
- Market trend data
- Prior showing feedback patterns

Logic Applied:

Analyze feedback patterns:

- Which features did they love/dislike?
- Price sensitivity indicators
- Location preference signals
- Timeline urgency level

Search MLS for matches:

- Required features from "love" feedback
- Exclude dealbreaker features
- Price range adjusted by feedback
- Neighborhoods showing interest

Prioritize recommendations:

- Properties listed in last 7 days (hasn't seen yet)
- Price reductions matching their range
- Properties with high-priority features

Actions Taken:

1. Send post-showing feedback form via text (2 hours after showing)

2. Compile feedback responses automatically

3. Generate AI summary identifying patterns

4. Search MLS for 3-5 highly-targeted new matches

5. Draft personalized email: "Based on your feedback about [specific feature], here are three properties that better match..."

6. Queue for agent review

7. After approval, send to client with showing availability

Quality Checkpoint:

Agent reviews AI-identified patterns and proposed properties before sending recommendations. This ensures local knowledge (e.g., "that neighborhood floods in heavy rain") prevents bad recommendations.

The Competitive Advantage:

Your competitors send generic "here are some new listings" emails. Your system sends "Based on your comment that the kitchen in yesterday's property felt too dark, here are three homes with kitchens featuring large windows and natural light" emails. Which agent seems more attentive?

Stage 3: Transaction Management

The Manual Approach:

Contract acceptance triggers a mental checklist. You create your timeline. You email all parties. You track deadlines manually. You send status updates weekly. You worry you're forgetting something. Each transaction requires constant mental overhead.

The Automated System:

Trigger: Contract executed and details entered into transaction management system

Data Sources Accessed:

- Contract terms and all critical dates
- State-specific timeline requirements
- Service provider contact information
- Client communication preferences
- Template status update language

Logic Applied:

Generate transaction timeline:

- Extract all contractual deadlines
- Add jurisdiction-specific requirements
- Include standard milestones (title order, inspection, appraisal, etc.)
- Build critical path (what must happen before what)
- Set automated reminders for all parties

Determine communication cadence:

- Daily updates during critical periods (inspection, appraisal)
- Bi-weekly updates during slower periods
- Immediate alerts for issues requiring attention

Actions Taken:

1. Generate complete transaction timeline document
2. Send to all parties (buyer, seller, lenders, attorneys, etc.)
3. Schedule automated reminder emails:
   - 7 days before each deadline
   - 3 days before each deadline
   - Day before each deadline
4. Generate weekly status update email with completed milestones and upcoming items
5. Queue for agent review and personalization

6. Track completion of each milestone

7. Alert agent immediately if deadline approaching without completion

Quality Checkpoint:

Agent reviews weekly status updates before sending, adding personal notes about conversations or concerns. Agent receives escalation alerts for any missed deadlines requiring immediate intervention.

What This Eliminates:

The 2 AM panic of "Did I order the title work?" The Friday afternoon realization you forgot to remind the buyer about their inspection contingency deadline. The mental load of tracking 6-8 simultaneous transactions with 150+ total action items.

Stage 4: Post-Closing Relationship

The Manual Approach:

You intend to stay in touch. You make a note to check in quarterly. Life gets busy. Six months pass. You see their name and think "I should reach out." Another month passes. Two years later, they list with someone else.

The Automated System:

Trigger: Closing recorded in CRM

Data Sources Accessed:

- Client transaction details
- Home purchase anniversary date
- Market data for their neighborhood
- Seasonal maintenance calendar
- Referral performance history

Logic Applied:

Build 24-month relationship sequence:

Month 1: Congratulations + home maintenance guide

Month 3: Neighborhood market update + "how's the house?"

Month 6: Seasonal maintenance reminders

Month 9: Holiday card + referral request

Month 12: Home value update + purchase anniversary celebration

Month 15: Market trends in their area

Month 18: Referral incentive reminder

Month 21: Home improvement ROI guidance

Month 24: Comprehensive market analysis + "ready to upgrade?"

Adjust based on engagement:

- High engagement: Increase communication frequency
- Low engagement: Reduce to quarterly touchpoints
- Zero engagement: Flag for personal call

Actions Taken:

1. Automatically generate personalized messages for each milestone
2. Pull current market data for their specific property/neighborhood
3. Queue emails for agent review and personalization
4. Send approved messages on schedule
5. Track engagement (opens, clicks, responses)
6. Adjust future communications based on engagement patterns
7. Alert agent when clients show signs of readiness (high engagement, life changes, market conditions)

Quality Checkpoint:

Agent personalizes automated messages with specific memories or details from transaction before sending. Agent reaches out personally for annual anniversary and major life events (births, marriages, job changes detected via social media).

The Referral Multiplier:

Consistent, valuable communication = top-of-mind awareness = referrals. Agents with post-closing automation report 40-60% of business from repeat/referral clients vs. 20-30% for agents without systematic follow-up.

⟋ Future-Ready Checkpoint: From Workflow Design to Workflow Supervision

The client journey automation you're building now (triggers, data sources, decision logic, actions, checkpoints) is training you to think systemically. In 2028, when agentic AI can manage these workflows autonomously with minimal intervention, your ability to design comprehensive client experiences will be the differentiator. You're not just learning tools. You're learning to architect automated businesses.

Advanced Integration: The GTM Engineering Approach

For agents ready to move beyond basic automation, **Go-To-Market (GTM) Engineering** represents the frontier of systematic business development. This approach, emerging from the B2B sales world but highly applicable to real estate, involves building sophisticated lead generation systems that operate continuously without manual prospecting.

What is GTM Engineering?

GTM Engineering is the discipline of connecting data sources, enrichment platforms, and outreach tools into automated systems that identify ideal prospects, gather context about them, and initiate personalized contact at the moment they show buying or selling intent signals.

Traditional prospecting:

You decide to target luxury condo owners. You pull a list. You send generic mailers. You hope someone responds. Response rate: 0.5-2%.

GTM Engineering approach:

Your system continuously monitors public data for intent signals (recent permits filed, property tax appeals, LLC formations, new business licenses, executive job changes). When a signal appears matching your ideal client profile, the system automatically researches that person, builds context, generates personalized outreach referencing their specific situation, and delivers it within hours of the intent signal appearing.

Response rate: 8-15%.

The difference isn't the message. It's the timing and relevance.

The Three-Tool Foundation

Modern GTM systems typically combine three types of platforms:

1. Sourcing/Contact Database (Finding prospects)

Examples: Apollo.io, ZoomInfo, LinkedIn Sales Navigator

Purpose: Identifying people matching your ideal client profile

Real estate application: Finding property owners, investors, developers, relocating executives

2. Enrichment Platform (Adding context)

Examples: Clay.com, Clearbit, Apify

Purpose: Layering additional data onto prospects from 50-100+ sources

Real estate application: Property ownership history, permit filings, business expansions, hiring signals, local connections

3. Orchestration Tool (Connecting everything)

Examples: Make.com, Zapier, n8n

Purpose: Building the logic that connects data sources to actions

Real estate application: "When X happens, check Y data, then do Z"

Story Behind the System: Investor Outreach Automation

Let me illustrate with a complete example targeting a specific niche: owners of small multifamily properties (2-4 units) who recently filed significant renovation permits.

Why this niche?

Renovation permits signal active management, capital deployment, and potential intent to improve for sale or refinance. These owners are in "action mode," making them dramatically more receptive to contact than random property owners.

The System Architecture:

Data Source: County building permit database (most counties publish this as open data or RSS feed)

Trigger: New permit filed for property classified as "multifamily" with valuation over $50K

Enrichment Workflow:

Step 1: Permit detected → Extract property address and owner name

Step 2: Look up property details:

- Purchase date and price
- Current estimated value
- Mortgage information (if public record)
- Number of units
- Rental rate estimates for area

Step 3: Find owner contact information:

- Search contact database for owner name
- Pull email, phone, LinkedIn profile
- Identify if owner is individual, LLC, or management company

- Find decision-maker if company

Step 4: Gather context:

- Owner's other property holdings
- Recent sales or purchases
- Local market trends for similar properties
- Typical cap rates in that submarket

Step 5: Generate personalized outreach:

"Hi [Name],

I noticed you recently filed a renovation permit for your property at [Address].

Having worked with several owners of 2-4 unit properties in [Neighborhood], I know that strategic improvements typically increase rental income by 15-25% in that area.

Given the current market conditions in [Neighborhood] (cap rates around X%, strong demand for renovated units), I'd love to share some recent comparable sales that

might inform your strategy, whether you're planning to hold long-term or considering a sale after renovations.

Would a 15-minute call this week be helpful? I have data on recent investor sales in your area that most owners find valuable.

[Your Name]"

Step 6: Route for approval:

- High-value prospects (properties worth $500K+): Queue for manual review
- Standard prospects: Send automatically from pre-approved template
- Log all activity in CRM

What Makes This Powerful:

1. **Intent Signal:** You're contacting them when they're actively engaged with their property, not randomly.

2. **Personalization:** The message references their specific property, their specific action, and your specific expertise in their situation.

3. **Value Proposition:** You're offering useful information (comps, market data), not asking for their business immediately.

4. **Timing:** Outreach happens within 24-48 hours of permit filing, while their intent is fresh.

5. **Scale:** The system runs continuously. One permit filing triggers the entire workflow automatically. You can operate this across your entire market without manual prospecting time.

Tools Involved:

- **Data Scraping:** Apify or custom scraper monitoring county permit database
- **Enrichment:** Clay.com pulling property data, ownership history, and contact information
- **AI Generation:** ChatGPT or Claude API generating personalized message
- **Orchestration:** Make.com connecting all components
- **CRM Integration:** Automatically logged for follow-up tracking

Your Role:

1. **Design:** Define ideal client profile and intent signals
2. **Approval:** Review high-value outreach before sending
3. **Follow-up:** Respond to replies and take conversations forward
4. **Optimization:** Monitor response rates and refine targeting/messaging

Learning Curve Reality:

Building this system requires 10-20 hours of learning and implementation. It's not plug-and-play. But once running, it operates indefinitely with minimal maintenance. The ROI calculation:

- Time to build: 15 hours
- Cost: $100-150/month in tool subscriptions

- Result: 10-15 qualified conversations per month that wouldn't have happened otherwise
- Close rate: 10-15% (1-2 deals per month)
- Value per deal: $8,000-15,000

Even one additional closing annually justifies the entire investment.

Identifying Your GTM Opportunities

Not everyone should build investor outreach systems. The key is identifying intent signals relevant to your niche:

For Luxury Home Specialists:

- Intent signal: Executive job changes to your area (LinkedIn, company announcements)
- Intent signal: High-value home sales (they'll likely re-buy locally)
- Intent signal: New business formations by affluent entrepreneurs

For First-Time Buyer Specialists:

- Intent signal: Apartment renters searching home listings (website tracking)
- Intent signal: Life events (engagements, births via social media)
- Intent signal: First-time home buyer seminar registrations (partnerships)

For Commercial Specialists:

- Intent signal: Business expansion announcements (hiring, funding, new locations)
- Intent signal: Lease expiration dates (public record in some jurisdictions)
- Intent signal: New LLC formations (expansion or investment activity)

For Relocation Specialists:

- Intent signal: Out-of-state license plate registrations (DMV data where accessible)
- Intent signal: Corporate relocation announcements

- Intent signal: Online searches for "moving to [your city]" from specific geo-locations

The GTM Engineering Mindset:

Stop thinking "How do I find clients?" Start thinking "What actions do my ideal clients take right before they need me, and how can I systematically detect and respond to those actions?"

Just-in-Time Learning: "Public Data is Everywhere"

County records, business filings, permit databases, LinkedIn activity, company announcements, social media, news articles, RSS feeds. Your ideal clients are leaving digital signals constantly. The question isn't whether data exists. It's whether you're systematically monitoring and responding to it. GTM Engineering is simply the discipline of doing systematically what successful agents have always done intuitively.

The Platform Stack: Understanding What to Look For

Rather than recommend specific tools (which change constantly), let me teach you how to evaluate platforms for your workflow needs.

What to Look for in CRM Systems

Must-Have Capabilities:

- **API access:** Can other tools connect to it programmatically?
- **Custom fields:** Can you track non-standard data points?
- **Automation/workflows:** Can you build if-then logic within the CRM?
- **Tagging system:** Can you segment contacts flexibly?
- **Activity logging:** Does it track all client interactions automatically?

Red Flags:

- Proprietary data formats that don't export cleanly
- No API or webhook capabilities
- Rigid structure that doesn't adapt to your workflow
- Poor mobile experience (you're often on-the-go)

Evaluate by asking: "If I wanted to automatically update this CRM when a lead engages with my website, is that possible without manually entering data?"

What to Look for in Automation Platforms

Must-Have Capabilities:

- **Visual workflow builder:** Can you see the logic flow?
- **Error handling:** What happens when something fails?
- **Conditional logic:** Can you build complex if-then-else scenarios?
- **Multiple integrations:** Does it connect to your existing tools?
- **Testing mode:** Can you trial workflows before going live?

Red Flags:

- Requires coding knowledge for basic automation
- Limited number of allowed workflows or actions
- Unreliable execution (workflows fail frequently)
- No clear documentation or support

Evaluate by asking: "Can I build a workflow where a form submission triggers data lookup, decision logic, and personalized email without writing code?"

What to Look for in Data Enrichment Tools

Must-Have Capabilities:

- **Multiple data sources:** Not just one database

- **Flexible queries:** Can you search by various criteria?
- **Data freshness:** Is information current or stale?
- **Export options:** Can you use the data elsewhere?
- **Compliance:** Do they follow data privacy regulations?

Red Flags:

- Inaccurate or outdated data
- Single data source (vulnerability)
- Expensive per-record pricing
- Unclear data sourcing (legal risk)

Evaluate by asking: "Can this tool tell me not just contact information, but context about why this person might need my services now?"

What to Look for in AI Generation Tools

Must-Have Capabilities:

- **API access:** Can you call it programmatically from workflows?
- **Customizable outputs:** Can you control tone, length, format?
- **Context retention:** Does it remember earlier in the conversation?
- **Variable insertion:** Can you dynamically include specific data?
- **Quality consistency:** Does it reliably produce good results?

Red Flags:

- Unpredictable output quality
- Can't customize for your brand voice
- No ability to save and reuse successful prompts
- Rate limiting that prevents workflow use

Evaluate by asking: "Can I send this tool data about a person and have it generate a personalized message that sounds like me and references their specific situation?"

The Minimum Viable Stack

You don't need every tool. Start with:

1. **Your existing CRM** (ensure it has automation capabilities)

2. **One automation platform** (Make.com, Zapier, or similar)

3. **One AI tool with API access** (ChatGPT, Claude, or Perplexity)

That's sufficient to build 80% of valuable workflows. Add specialized tools only when you've identified specific gaps.

Future-Ready Checkpoint: Platform-Agnostic Thinking

Tool names will change. Platforms will merge or disappear. But the underlying capabilities (data enrichment, workflow orchestration, AI generation, contact management) remain constant. By learning to evaluate platforms based on capabilities rather than brand names, you build resilient knowledge that survives inevitable market disruption.

Ethical Guardrails for Advanced Automation

As your systems become more sophisticated, ethical considerations intensify. Automation at scale amplifies both good practices and bad ones.

The Three Non-Negotiables

1. Fair Housing Compliance in All Automated Systems

Any system that targets prospects, scores leads, or personalizes messaging MUST be audited for Fair Housing compliance.

The Risk:

Algorithms can perpetuate bias even when you don't intend it. If your lead scoring gives higher priority to prospects from certain neighborhoods, or your targeting excludes protected classes, you're potentially violating Fair Housing laws even if your system does it automatically.

The Solution:

- Explicitly instruct all AI systems to avoid referencing protected classes
- Base targeting on behavior and intent signals, never demographics
- Audit outputs regularly for biased language or patterns
- Document your compliance protocols
- Consult with legal counsel on automated marketing systems

Example of compliant targeting: "Property owners who filed renovation permits in the last 30 days" (behavior-based)

Example of non-compliant targeting: "Property owners in neighborhoods with high minority populations" (demographic-based)

2. Data Privacy and Confidentiality

Public data is fair game. Confidential client information is not.

Never input into public AI tools:

- Client financial information
- Confidential transaction details
- Social Security numbers or sensitive personal data
- Proprietary contracts or negotiation details
- Information learned under confidentiality agreements

Always verify:

- Data source legitimacy (is this publicly available information?)
- Tool vendor security practices (how is data stored and used?)
- Your legal obligations (what did you promise clients?)

Use enterprise tools with proper security: If you're automating workflows involving sensitive client data, use enterprise-grade tools with proper data protection agreements, not free consumer tools.

3. Human Authority and Oversight

Automation should enhance your judgment, not replace it.

Mandatory human checkpoints for:

- Final pricing recommendations
- Contract terms and legal documents
- Unusual situations or edge cases
- High-value client communications
- Anything that could significantly impact a client's outcome

The Rule: If you wouldn't delegate a decision to an unlicensed assistant without review, don't delegate it to an AI system without review.

Transparency with Clients

You don't need to disclose every AI tool you use, but you should be transparent about automated systems that affect clients.

Recommended Disclosure Language:

"I use advanced technology and automation to enhance my service in several ways: faster responses to your questions, more comprehensive market analysis, and consistent communication throughout your transaction. These tools assist me, but I personally review all advice and maintain full responsibility for the guidance I provide you."

When to disclose specifically:

- Virtual staging in photos (always)
- Automated property valuation tools (when presenting pricing)
- AI-assisted content creation (if client asks)
- Automated communication systems (if they prefer not to receive them)

The Standard: Would a reasonable client feel misled if they learned about your AI use later? If yes, disclose proactively.

The Escalation Protocol

Build safeguards into every automated system:

Define escalation triggers:

- Unusual client language (anger, confusion, distress)
- Transaction complications requiring judgment
- Data discrepancies or errors
- Requests outside normal parameters
- Time-sensitive urgent matters

When triggers activate:

- Pause automation immediately
- Alert human agent for intervention
- Provide context: "System escalated because [specific trigger]"
- Ensure smooth handoff from automated to personal attention

Example:

Your automated post-closing check-in email generates a response: "The septic system failed two days after closing and the inspector says it was pre-existing damage."

Your system should:

1. Flag this as urgent (keyword: "failed," "inspector," "pre-existing")
2. Pause all automated communications with this client
3. Alert you immediately with the full context
4. Move client into "needs immediate personal attention" queue
5. Log the escalation for quality review

Never let an automated system:

- Respond to legal issues or complaints
- Make promises or commitments on your behalf
- Handle complex negotiations
- Address crisis situations

Automation handles routine excellence. Humans handle exceptions and complexity.

ⅡⅠ Life Margin Impact: The Paradox of Advanced Automation

Here's what agents implementing advanced automation report: they work fewer hours while earning more income, but they also report feeling more connected to clients, not less. Why? Because automation handles the transactional noise, allowing them to be fully present for meaningful conversations. They're not distracted by "did I remember to..." because systems handle remembering. They're not rushed through consultations because they've reclaimed 10-15 hours weekly. The result isn't mechanized service. It's more human service, enabled by systems that handle the mechanical parts.

Starting Your Advanced Automation Journey

Don't attempt to build everything at once. Here's your realistic implementation path:

Month 1: Assess and Design

Week 1: Audit your current workflows

- Identify your three biggest manual bottlenecks
- Document the steps involved in each
- Determine what data you need at each step
- Identify where human judgment is essential

Week 2: Select one workflow to automate

- Choose based on frequency, time consumption, potential impact
- Map out the ideal automated workflow on paper
- Identify required tools and data sources
- Calculate projected time savings

Week 3: Research and select tools

- Identify 2-3 platform options for each capability needed

- Sign up for free trials
- Test each with sample data
- Select your stack

Week 4: Design your first workflow

- Map trigger, data sources, logic, actions, checkpoints
- Create flowchart or written documentation
- Review for ethical compliance and quality checkpoints
- Get feedback from peers or mentors if available

Month 2: Build and Test

Week 1: Build the basic workflow

- Connect tools and test data flow
- Build the logic step-by-step
- Test with sample data, not real clients
- Document any issues or limitations

Week 2: Refine and troubleshoot

- Address errors and edge cases
- Add proper error handling
- Build in escalation triggers
- Create backup plans for failures

Week 3: Pilot with real data

- Run alongside your manual process (don't rely on automation yet)
- Compare automated output to manual output
- Track any issues or inaccuracies
- Refine based on real-world performance

Week 4: Document and deploy

- Create written documentation of how the workflow operates
- Document troubleshooting steps for common issues
- Train team members if applicable
- Go live with automated system while monitoring closely

Month 3: Monitor and Optimize

Ongoing: Track performance metrics

- Time saved per use
- Error rate or issues encountered
- Client feedback (if client-facing)
- Financial impact (cost vs. value generated)

Adjust and improve:

- Refine logic based on edge cases discovered
- Optimize messaging based on response rates
- Expand to additional use cases
- Consider adding second workflow

Month 4+: Scale and Expand

Strategic expansion:

- Add complementary workflows that connect to your first
- Build client journey automation that spans multiple touchpoints
- Implement GTM Engineering for business development
- Continuously optimize based on performance data

The Realistic Timeline:

- First simple automation: 2-4 weeks
- First complex automated workflow: 4-8 weeks
- Fully automated client journey: 3-6 months
- Sophisticated GTM system: 6-12 months

Don't rush. One well-built, reliable workflow is infinitely more valuable than five half-working systems that require constant troubleshooting.

An AI-Amplified Agent in Action

Let me share the story of Marcus, a commercial real estate agent who built an advanced automation system over 18 months.

Starting Point (Month 0):

- 8-12 transactions annually
- 55-60 hour work weeks
- Constantly reactive, rarely proactive
- Inconsistent follow-up with past clients
- Manual prospecting when he had time (rarely)

First Automation (Month 2):

Transaction timeline automation. Every contract triggered automatic timeline generation, reminders to all parties, and weekly status updates to clients. Time saved: 3 hours per transaction. Mental load eliminated: significant.

Second Automation (Month 5):

Post-closing relationship system. Automated quarterly check-ins, market updates, and referral requests for past clients. Result: Referral rate increased from 18% to 34% of business within one year.

Third Automation (Month 9):

Lead enrichment workflow. New leads automatically researched, scored, and queued with complete dossiers. Response time dropped from hours to minutes. Lead conversion increased 28%.

Advanced System (Month 16):

GTM Engineering system targeting business owners expanding to second locations. Monitored new business licenses, commercial permits, and company growth signals. Generated 15-20 qualified conversations monthly that would never have happened through manual prospecting.

Results After 18 Months:

- Transactions: 8-12 annually → 18-22 annually
- Work hours: 55-60 hours → 42-45 hours
- Income: +140%
- Stress level: Dramatically reduced
- Client satisfaction: Measurably higher (post-transaction surveys)

- Life margin: Coaching kid's soccer, regular date nights, annual two-week vacation

Marcus's Reflection:

"I thought I was going to automate myself into a robot agent. Instead, I automated the robot parts so I could be more human. My clients get better service because I'm not drowning in administrative tasks. I'm present in conversations. I'm thinking strategically about their needs. And ironically, I'm building deeper relationships while working fewer hours. That's the amplified agent advantage."

That's what this chapter enables. Not mechanized service. Amplified humanity.

What You've Accomplished

By working through this chapter, you've:

✓ **Understood the shift from tools to systems** and why connected workflows multiply impact

✓ **Learned the five components** of every automated workflow (trigger, data, logic, actions, checkpoints)

✓ **Designed client journey automation** for acquisition, active engagement, transactions, and relationship maintenance

✓ **Explored GTM Engineering** for systematic business development using intent signals

✓ **Developed platform evaluation criteria** that survives changing tool landscapes

✓ **Established ethical guardrails** for compliant, responsible automation at scale

✓ **Created your implementation roadmap** for the next 3-6 months

More importantly, you've developed **systems thinking**. The ability to design workflows that operate without constant intervention while maintaining quality and ethics. This is the mindset that separates amplified agents from overwhelmed agents.

Transition to Chapter 7

You've built sophisticated systems. Workflows are running. Time is being reclaimed. Clients are experiencing seamless service. But one critical question remains:

How do you prove it's working?

Chapter 7 provides the frameworks for measuring impact, calculating true ROI, and demonstrating (to yourself and others) that your AI investment is delivering tangible business results. You'll learn to track not just time saved, but quality improved, revenue generated, and life margin reclaimed.

Chapter 7: Results & ROI

The Numbers That Changed Everything

Jamie didn't believe in tracking metrics. She was an agent, not an accountant. Her business ran on relationships and instinct. Numbers were for people who didn't understand that real estate was fundamentally about human connection.

Then she had the worst quarter of her career.

Four deals fell through. Six listings sat too long. Her closing rate dropped to barely half her previous average. She had no idea why. She just knew she was working harder than ever while earning less than the year before.

Her broker asked a simple question: "Do you know which activities are actually driving your closed transactions versus which ones just feel productive?"

Jamie didn't. She had no idea how many leads she was generating monthly. No clue what her conversion rate looked like. No tracking of which marketing efforts produced actual business versus which just consumed time and budget.

"Here's the thing," her broker said. "You can't improve what you're not measuring. And right now, you're flying blind."

So Jamie reluctantly started tracking five basic metrics. Not elaborate analytics. Just five numbers she reviewed weekly:

1. Lead generation (quantity and source)

2. Conversion rate (leads to clients)

3. Average time per transaction stage

4. Client satisfaction scores

5. Referral rate

Three months later, the data told a story her instincts had missed:

Her highest-converting leads came from past client referrals (42% conversion) and her educational workshop series (38% conversion). Her lowest-converting leads came from expensive Facebook ads (6% conversion) and open houses (9% conversion).

She was spending 60% of her marketing budget on the lowest-performing channels and almost nothing reinforcing her highest performers.

Armed with data, Jamie made changes. She cut paid advertising by 70%. She doubled down on past client relationship nurturing. She expanded her workshop series. She implemented systematic referral requests.

Six months later: conversion rate up 45%, average transaction time down 22%, client satisfaction scores at all-time highs, referral business doubled.

"I was wrong," Jamie told her broker. "Metrics aren't about being an accountant. They're about seeing clearly instead of guessing. They're about knowing what's working so you can do more of it and what isn't so you can stop wasting time on it."

That clarity is what this chapter gives you. Not boring spreadsheets. **Strategic vision powered by data.**

Reframing Metrics: Your Business Health Dashboard

Let's be honest: most agents don't track metrics consistently. It feels tedious. Administrative. A distraction from the "real work" of serving clients and closing deals.

But here's what successful agents understand that struggling agents don't: **metrics are your early warning system.**

Think of metrics like the dashboard in your car. You don't stare at it constantly. But you glance at it regularly to catch problems before they become crises:

- Fuel gauge dropping? You find a gas station before you're stranded.
- Check engine light? You address it before costly damage occurs.
- Speed creeping up? You adjust before getting pulled over.

Your business metrics serve the same purpose. They reveal:

- **Trends before they become disasters:** Lead generation dropping 15% monthly (catch it before pipeline dries up)
- **Hidden inefficiencies:** Spending 10 hours on tasks that generate zero revenue
- **What's actually working:** Which marketing produces buyers vs. which produces tire-kickers
- **When to pivot:** Market shifts requiring strategy adjustments
- **ROI on investments:** Whether that new tool or service is worth continuing

Metrics aren't about judgment. They're about clarity.

No one is grading your conversion rate. You're not competing with other agents' numbers. You're simply establishing visibility into your own business so you can make informed decisions instead of emotional guesses.

The Gamification Mindset

Here's a better way to think about metrics: you're leveling up your business.

Video games are addictive because they show constant feedback. You see your progress. You watch your stats improve. You unlock new capabilities. You level up.

Your business can work the same way. Track your metrics. Watch them improve. Celebrate milestones. Level up your capacity.

Level 1 Agent (Just starting with AI):

- Tracks basic time savings
- Monitors content output
- Reviews client feedback informally

Level 2 Agent (Building AI workflows):

- Tracks conversion rates by source
- Measures quality improvements
- Calculates financial ROI

Level 3 Agent (Advanced automation):

- Sophisticated funnel analysis
- Predictive trend identification
- Comprehensive business intelligence

You don't need Level 3 metrics on Day 1. Start where you are. Track what matters at your stage. Level up as your implementation matures.

📝 Future-Ready Checkpoint: Data-Driven Agentic AI

In the very near future, agentic AI systems will continuously monitor your business metrics and proactively suggest adjustments: "Your lead response time increased 35% over the last two weeks. Would you like me to implement faster automated acknowledgment?" Agents who understand metrics now will know how to evaluate these AI recommendations. Those who don't will accept suggestions blindly without understanding if they're beneficial.

The Five Metrics That Actually Matter

You could track 100 data points. Most would be noise. Focus on five that directly impact your business success and life margin goals.

Metric 1: Time Reclaimed (The Foundation)

What it measures: Hours saved per week through AI implementation

Why it matters: This is your life margin metric. Time saved either converts to earning capacity (more clients served) or quality of life (shorter work weeks, actual vacations).

How to track:

- Identify your three most time-consuming weekly tasks

- Time them manually for two weeks (establish baseline)
- Implement AI assistance
- Time them again for two weeks (measure new state)
- Calculate weekly hours saved

Example Tracking:

BASELINE (Manual):

- Email management: 8 hrs/week
- Listing prep: 6 hrs/week
- Social content creation: 5 hrs/week

TOTAL: 19 hrs/week

WITH AI (3 months in):

- Email management: 3 hrs/week (AI-drafted responses)
- Listing prep: 2.5 hrs/week (AI descriptions, auto-staging)
- Social content creation: 1.5 hrs/week (AI content generation)

TOTAL: 7 hrs/week

TIME RECLAIMED: 12 hrs/week = 624 hrs/year

What to do with this metric:

- If hours saved are minimal, you're not implementing effectively (Chapter 8 troubleshooting)
- If hours saved are significant, decide how to reinvest them (more clients? Better work-life balance?)
- Track this monthly to ensure gains don't erode over time

Level Up Milestone:

- Level 1: Saving 5+ hours weekly
- Level 2: Saving 10+ hours weekly
- Level 3: Saving 15+ hours weekly while maintaining or improving quality

---

Metric 2: Lead Conversion Rate (The Revenue Driver)

What it measures: Percentage of leads that become clients

Why it matters: This is your efficiency metric. More important than lead quantity is lead quality and your ability to convert them. AI should improve this by enabling faster response, better personalization, and consistent follow-up.

How to track:

Monthly Conversion Rate = (New Clients / Total Leads) × 100

Example:

40 leads generated in March

6 became clients

Conversion Rate: 15%

Track by source separately:

Lead Source	March Leads	Conversions	Rate
Past Client Referrals	8	4	50%
Educational Workshops	12	5	42%
Website Contact Form	15	3	20%
Open House Sign-ins	20	2	10%
Facebook Ads	25	1	4%

What this reveals:

- Your highest-quality lead sources (focus here)
- Your lowest-performing channels (reduce or eliminate)
- Whether AI follow-up is improving conversion (track pre/post implementation)

- Where to invest time and money for maximum return

What to do with this metric:

- Double down on channels with >30% conversion
- Improve or eliminate channels with <10% conversion
- A/B test AI-enhanced follow-up vs. manual to see impact
- If conversion drops after AI implementation, investigate why (quality issues? impersonal feel?)

Level Up Milestone:

- Level 1: Knowing your conversion rate (most agents don't)
- Level 2: Tracking by source and optimizing channel mix
- Level 3: Conversion rate improving 25%+ from AI implementation

---

Metric 3: Response Time (The Competitive Advantage)

What it measures: Time from lead inquiry to meaningful response

Why it matters: In real estate, speed kills competition. Studies show that responding to a lead within 5 minutes vs. 30 minutes can increase conversion by 100-400%. AI automation should dramatically accelerate your response time.

How to track:

Use your CRM or email timestamps:

Calculate weekly: Sum of (Response Time per Lead) / Total Leads

Example Week:

Lead 1: 12 minutes

Lead 2: 45 minutes

Lead 3: 8 minutes

Lead 4: 2 hours (120 minutes)

Lead 5: 22 minutes

Average: 41.4 minutes

Target benchmarks:

- Competitive: Under 15 minutes average
- Excellent: Under 5 minutes average
- AI-powered goal: Under 2 minutes for automated acknowledgment + personalized follow-up within 30 minutes

What to do with this metric:

- If response time is >1 hour: implement immediate lead notification automation
- If under 15 minutes: you're competitive, now optimize quality of response
- Track response time by time of day (evening/weekend leads often delayed; automate these)

The AI Enhancement:

Before AI: Average response time 3.5 hours (most during business hours, much slower after 5pm)

After AI:

- Automated acknowledgment: 90 seconds average
- AI-researched, personalized follow-up: 12 minutes average
- 24/7 consistency (no more weekend/evening delays)

Level Up Milestone:

- Level 1: Tracking response time consistently
- Level 2: Under 15 minutes average
- Level 3: Under 5 minutes with AI-automated first response, personalized human follow-up within 30 minutes

---

Metric 4: Client Satisfaction Score (The Sustainability Indicator)

What it measures: How clients rate their experience with you

Why it matters: This is your business sustainability metric. High satisfaction = referrals, repeat business, and reputation. AI should enhance satisfaction by improving consistency, responsiveness, and professionalism. If satisfaction drops after AI implementation, you're automating poorly.

How to track:

Simple post-transaction survey (3-5 questions):

"On a scale of 1-10, how would you rate:

1. Overall experience working with me

2. Responsiveness to your questions and concerns

3. Quality of guidance and advice provided

4. Likelihood you'd recommend me to friends/family

Optional: What could have been better?"

Calculate:

Average Score = Sum of all ratings / Total responses

Target: 8.5+ average across all questions

Red flag: Anything under 7.5 average

What to do with this metric:
- If scores drop after AI implementation: investigate (feels impersonal? quality issues?)
- If scores increase: document what's working and expand those practices
- Pay attention to qualitative comments; they often reveal insights numbers don't
- Track "likelihood to recommend" specifically; this predicts referral business

What NAR research reveals:

According to the 2024 NAR Profile of Home Buyers and Sellers, when clients were asked what benefits they received from their agent, the most cited were:

- Helping buyer understand the process: 61%
- Pointing out unnoticed features/faults with property: 58%
- Negotiating better sales contract terms: 46%
- Providing a better list of service providers: 46%
- Improving buyer's knowledge of search areas: 45%

The AI Connection: Notice these are all human-judgment activities where AI serves as a support tool, not a replacement. Clients value agents who educate, guide, notice details, and negotiate skillfully. AI should free you to focus more attention on these high-value activities.

What clients prioritize when choosing an agent:

For Buyers:

- Agent's experience: 21%
- Honesty and trustworthiness: 19%
- Reputation: 15%
- Agent was friend/family: 12%

For Sellers:

- Reputation of agent: 35%
- Agent is honest and trustworthy: 21%
- Agent is friend or family member: 16%
- Agent's knowledge of the neighborhood: 10%

The takeaway: Clients choose agents based on trust, reputation, and expertise. Your AI implementation should support these qualities, never undermine them. If satisfaction metrics decline after AI adoption, you're likely over-automating or depersonalizing.

Level Up Milestone:

- Level 1: Collecting client satisfaction data consistently
- Level 2: Maintaining 8.5+ average while implementing AI
- Level 3: Satisfaction scores improving because AI enables better service quality and attention

---

Metric 5: ROI on AI Investment (The Business Case)

What it measures: Financial return on your AI tool subscriptions and implementation time

Why it matters: This justifies continued investment and guides budget decisions. Positive ROI means keep going. Negative ROI means troubleshoot or pivot.

How to track:

Monthly AI Costs:

- Tool subscriptions: $_____
- Implementation time × your hourly value: $_____

TOTAL MONTHLY INVESTMENT: $_____

Monthly Value Generated:

- Time saved × hourly value: $_____
- Additional transactions enabled: $_____
- Cost avoided (VA, services you no longer need): $_____

TOTAL MONTHLY VALUE: $_____

Monthly ROI = ((Value - Investment) / Investment) × 100

Detailed Example:

COSTS (Monthly):

- ChatGPT Plus: $20
- Canva Pro: $15
- Virtual staging: $39
- Make.com automation: $29
- Total subscriptions: $103
- Implementation time: 8 hrs × $75/hr = $600 (first month only)

TOTAL MONTH 1: $703

VALUE GENERATED (Monthly):

- Time saved: 12 hrs/week × 4 weeks = 48 hrs × $75/hr = $3,600
- Additional client capacity: Closed 1 extra transaction = $8,000
- Avoided costs: No longer paying VA $500/month

TOTAL VALUE: $12,100

Month 1 ROI: (($12,100 - $703) / $703) × 100 = 1,621%

Ongoing ROI (after implementation): (($12,100 - $103) / $103) × 100 = 11,589%

Realistic expectations:

- **First 3 months:** ROI may be modest or negative (learning curve, implementation time)
- **Months 4-6:** ROI should become clearly positive (500-1000%+ typical)
- **After 6 months:** ROI compounds as efficiency improves and capacity increases

What to do with this metric:

- If ROI is negative after 6 months: troubleshoot implementation or reconsider approach
- If ROI is positive but modest (<500%): identify bottlenecks limiting impact
- If ROI is strong (>1000%): expand implementation to additional workflows
- Use this data to justify additional AI investments or team adoption

Level Up Milestone:

- Level 1: Breaking even (value = cost) within 3 months
- Level 2: 500%+ ROI within 6 months
- Level 3: 1000%+ ROI with measurable capacity increase

The Weekly Scorecard: Making Measurement Effortless

The biggest reason agents don't track metrics isn't laziness. It's that tracking feels like a separate job on top of their already-overwhelming workload.

Solution: **The Weekly Scorecard.** Five minutes every Friday. Five numbers. That's it.

Your Weekly Scorecard Template

Create a simple spreadsheet with these columns:

Week Of	Time Saved (hrs)	New Leads	Conver- sions	Avg Response Time	Satisfaction (Surveys)
12/16	11	23	2	18 min	-
12/23	13	19	3	12 min	9.2 (2 responses)
12/30	10	14	1	9 min	-

How to complete it in 5 minutes:

1. **Time Saved:** Rough estimate (you timed baseline, now estimate how much faster you are)

2. **New Leads:** Pull from CRM or lead management system

3. **Conversions:** Count new clients signed this week

4. **Response Time:** Check 3-5 recent lead responses, average them

5. **Satisfaction:** Log any survey responses received (even if from deals that closed weeks ago)

The Power of This Simple System:

- **Trends become visible:** "My lead count is dropping for three weeks straight" (early warning)
- **Improvements are validated:** "Response time cut in half since automation went live"
- **Problems are caught early:** "Conversions dropped after changing my email template"
- **Wins are celebrated:** "12 hours saved weekly! I'm Level 2 now."

Monthly Review (15 minutes):

Once a month, look at your four weekly scorecards together:

- Calculate monthly averages
- Identify trends (improving? declining? flat?)
- Celebrate improvements
- Flag concerns for troubleshooting (Chapter 8)
- Adjust goals for next month

Advanced Metrics for Level 2-3 Agents

Once your basic scorecard is running smoothly (3-6 months), you may want to track additional insights:

Funnel Conversion Analysis

Track conversion at each stage (not just lead to client):

100 Leads (Top of Funnel)
↓
45 Engaged (responded to outreach) - 45% engagement rate
↓
20 Qualified (met with/spoke to) - 44% qualification rate
↓
12 Active Clients - 60% conversion from qualified
↓
8 Closed Transactions - 67% closing rate

What this reveals:

- Where leads drop off (focus improvement here)
- Which stage has lowest conversion (biggest opportunity)
- Impact of AI at different stages

Example insight: "My engagement rate is great (45%) but qualification rate is weak (44%). I'm attracting interest but not the right prospects. Need to adjust targeting or qualification process."

Cost Per Acquisition by Channel

Track what you spend per channel vs. clients acquired:

Channel	Monthly Spend	Clients Acquired	Cost Per Client
Facebook Ads	$800	1	$800
Past Client Events	$300	4	$75
Educational Workshops	$200	3	$67
SEO/Website	$150	2	$75

What this reveals: Past client nurturing and educational content have 10-12x better ROI than paid advertising for this agent. Shift budget accordingly.

Time Allocation Audit

Track where your hours go:

Weekly time log (one week per quarter):

Activity Category	Hours Spent	% of Total
Client meetings/showings	18	36%
Lead generation/follow-up	8	16%
Transaction coordination	12	24%
Marketing/content creation	6	12%
Administrative tasks	6	12%

What this reveals:

- Activities consuming disproportionate time
- Whether time allocation aligns with revenue generation
- Best automation targets (high time/low revenue activities)

After AI implementation, retrack:

Activity Category	Hours Spent	% of Total	Change
Client meetings/showings	20	50%	+2 hrs
Lead generation/follow-up	8	20%	0
Transaction coordination	6	15%	-6 hrs
Marketing/content creation	3	7.5%	-3 hrs
Administrative tasks	3	7.5%	-3 hrs

Goal: More time on revenue-generating activities (client meetings, qualified lead generation), less on administrative overhead.

The Metrics-to-Action Framework

Data without action is just trivia. Here's how to translate metrics into business improvements:

The Traffic Light System

For each metric, establish three zones:

Green Zone (Exceeding Target):

- Celebrate and document what's working
- Expand this approach to other areas
- Maintain current practices

Yellow Zone (At Target, Not Improving):

- Acceptable but not growing
- Look for incremental improvements
- Monitor for decline

Red Zone (Below Target):

- Requires immediate attention
- Investigate root causes
- Implement corrective action
- Track weekly until improved

Example:

Response Time Metric:

- Green: Under 10 minutes average
- Yellow: 10-30 minutes average
- Red: Over 30 minutes average

Your current average: 45 minutes (Red Zone)

Action Plan:

1. Implement automated lead acknowledgment (Chapter 6 workflows)

2. Set up mobile notifications for new leads

3. Create response templates for common inquiries

4. Track daily for two weeks

5. Target: Get to yellow zone within 2 weeks, green zone within 4 weeks

---

The "So What?" Test

For every metric you track, ask: "So what does this number mean I should DO differently?"

If you can't answer that question, you're tracking vanity metrics, not actionable data.

Example:

Metric: "I have 1,247 Instagram followers."

So what? "I should... uh... get more?" (Vanity metric. No clear action.)

Better metric: "My Instagram posts average 85 engagements and generate 2-3 leads monthly."

So what? "Instagram delivers 4% of my monthly leads for about 2 hours of effort. ROI is positive but modest. Maintain current effort but don't expand." (Actionable.)

Even better: "Instagram posts featuring client testimonials generate 3x more leads than property photos."

So what? "Shift Instagram content mix to 60% testimonial-based stories, 40% property content." (Specific action driven by data.)

Creating Your Personal Measurement System

Don't try to track everything. Build a simple system that fits your business model:

Step 1: Choose Your Starting Metrics (Week 1)

Select 3-5 metrics from the five core options based on your current priorities:

If your main goal is reclaiming time/life margin:

- Time Reclaimed
- Response Time
- Client Satisfaction

If your main goal is growing revenue:

- Lead Conversion Rate
- ROI on AI Investment
- Response Time

If your main goal is improving quality:

- Client Satisfaction
- Response Time
- Time Reclaimed

Step 2: Establish Baselines (Weeks 2-3)

Before implementing or expanding AI:

- Track your chosen metrics for 2 weeks manually
- Document current state honestly
- Calculate averages
- This is your baseline for measuring improvement

Step 3: Set Realistic Targets (Week 4)

For each metric, set achievable 90-day targets:

Example:

Current baseline → 90-day target

- Time Saved: 0 hrs/week → 8 hrs/week
- Response Time: 3.5 hrs average → 20 min average

- Conversion Rate: 11% → 15%
- Client Satisfaction: 7.8 average → 8.5 average
- ROI: $0 → 500%

Step 4: Implement and Track (Ongoing)

- Use your Weekly Scorecard (5 min every Friday)
- Monthly review and adjustment (15 min monthly)
- Quarterly deep dive (1 hour quarterly to audit time allocation and reassess targets)

Step 5: Tie Metrics to Chapter 8 Troubleshooting

Your metrics will tell you when something needs fixing:

Red flag patterns that require troubleshooting:

- Time saved decreasing over time → AI workflows breaking down or being abandoned
- Conversion rate dropping after AI implementation → Quality or personalization issues
- Response time increasing → Automation not working or turned off
- Satisfaction scores declining → Over-automation or impersonal service
- ROI negative after 6 months → Wrong tools, poor implementation, or unrealistic value calculation

When you spot these patterns, Chapter 8 provides the diagnostic frameworks to identify root causes and implement solutions.

Metrics are your early warning system. They tell you when to troubleshoot before problems become crises.

How Measuring ROI Changed Everything

Let's consider David, a real estate agent who was deeply skeptical about both AI and tracking metrics.

In his 50s, David felt he had successfully built his business on relationships, local knowledge, and hustle. He didn't think he needed algorithms and spreadsheets.

But his business had plateaued. Same revenue for three years. Longer hours every year. More competition from younger, tech-savvy agents.

His broker convinced him to try two things for 90 days:

1. Track five simple metrics weekly

2. Implement basic AI tools for email and listing prep

David's 90-Day Metrics Journey:

Baseline (Week 1):

- Time Reclaimed: 0 hrs (not using AI yet)
- New Leads: 18/week average
- Conversion Rate: 14%
- Response Time: 6.5 hours average
- Satisfaction Score: 8.3 average

Week 6 (Early Implementation):

- Time Reclaimed: 4 hrs/week
- New Leads: 19/week
- Conversion Rate: 16%
- Response Time: 2.1 hours average
- Satisfaction Score: 8.5 average

David's response time improvement was dramatic. He implemented automated lead acknowledgment and AI-drafted follow-ups that he'd review and personalize. Prospects were amazed at how quickly he got back to them, even at 9 PM or on weekends.

Week 12 (Fully Adopted):

- Time Reclaimed: 9 hrs/week

- New Leads: 22/week (had capacity for more outreach)
- Conversion Rate: 19%
- Response Time: 35 minutes average
- Satisfaction Score: 8.7 average
- ROI Calculation: 847%

What the numbers revealed:

- Response time was THE major factor in his conversion improvement
- He'd been losing deals simply by being slower than competitors
- The 9 hours reclaimed weekly allowed him to personally call 15-20 prospects he previously couldn't reach
- His satisfaction scores improved because he was less rushed and more present in client interactions

David's conclusion:

"I was wrong about metrics. They showed me exactly where AI could help most (response speed) and proved it was working (conversion rate up 36%). Now I review my scorecard every Friday. Takes five minutes. Tells me everything I need to know about whether my business is healthy or needs attention."

The transformation:

- Year 1 post-implementation: Revenue up 28%
- Year 2: Revenue up another 19%
- Work hours: Down from 55-60/week to 45/week
- Closed deals: Up from 14/year to 21/year

"The metrics didn't replace relationships and local knowledge," David said. "They gave me the clarity to amplify what was already working and fix what was broken."

What You've Accomplished

By working through this chapter, you've:

✓ **Reframed metrics** from boring data tracking to strategic vision and leveling up

✓ **Identified the five core metrics** that drive real estate business success

✓ **Created a simple Weekly Scorecard** that takes 5 minutes to maintain

✓ **Learned the Traffic Light System** for translating metrics into action

✓ **Understood what NAR research says** about what clients value

✓ **Built your personal measurement system** tied to your specific goals

✓ **Established the foundation for Chapter 8** troubleshooting with baseline metrics

More importantly, you've **gained visibility into your business** that most agents don't have. You're no longer guessing about what's working. You're no longer flying blind through market changes. You're making data-informed decisions that compound over time.

You're not just an agent with AI tools. You're an amplified agent with strategic clarity.

Transition to Chapter 8

You've built AI systems (Chapters 5-6). You're measuring their impact (Chapter 7). But what happens when something isn't working?

When your conversion rate drops after implementing automation. When client feedback suggests your communications feel impersonal. When AI outputs need constant editing. When tools don't integrate smoothly. When ROI stays stubbornly low despite your best efforts.

That's when measurement becomes diagnosis.

Chapter 8 takes your metrics data and transforms it into troubleshooting intelligence. You'll learn to:

- Identify root causes (not just symptoms)
- Recognize patterns indicating specific problems
- Apply targeted solutions to common obstacles
- Refine your AI systems for continuous improvement
- Know when to adjust vs. when to pivot completely

Your metrics tell you WHAT is happening. Chapter 8 teaches you WHY and HOW to fix it.

Every successful AI implementation encounters obstacles. The difference between agents who thrive and agents who abandon AI entirely is having a systematic framework for diagnosing and solving problems.

Let's build yours.

Chapter 8: Troubleshooting & Refinement

When Perfect Plans Hit Reality

Rachel had done everything right. She'd read the guides. Watched the tutorials. Set up her AI workflows carefully. Her automated lead follow-up system was supposed to revolutionize her business.

Three weeks in, her conversion rate dropped by 30%.

Leads were responding to her automated emails with one-word answers. Two prospects specifically mentioned that her communications felt "robotic." One wrote: "Are you even reading these, or is this all automated?"

Rachel panicked. Maybe AI wasn't right for her business. Maybe the skeptics were right. Maybe she should abandon the whole thing and go back to manual everything.

Then her broker asked a simple diagnostic question: "What changed when your conversion rate dropped?"

Rachel pulled her Chapter 7 metrics. The pattern was clear:

Week 1-3 (before automation): 18% conversion rate, personalized manual follow-ups

Week 4-6 (after automation): 12% conversion rate, AI-generated follow-ups

"Show me one of your automated emails," her broker said.

Rachel pulled up an example:

Hello [First Name],

Thank you for your inquiry about properties in [City].

I have extensive experience in the [City] real estate market and would love to help you find your ideal home. I have several properties that may interest you based on your criteria.

When would be a good time for us to discuss your needs?

Best regards,

Rachel [Last Name]

[Title] at [Brokerage]

Her broker pointed at the screen. "This reads like a form letter. Where's your voice? Where's the specific reference to what they inquired about? Where's the value you're offering beyond 'let's schedule a call'?"

Rachel had automated the process but not personalized the content. She'd built the workflow correctly but forgotten that AI amplifies whatever you give it. Generic input creates generic output at scale.

The fix took two hours:

She rebuilt her automated follow-up to pull specific data from each lead's inquiry, reference the actual properties they viewed, and include one genuinely valuable insight personalized to their situation. She added her actual voice by providing AI with examples of her best manual emails and instructing it to match that tone.

Week 7-9 (after refinement): 24% conversion rate. Higher than her manual baseline.

The difference wasn't abandoning automation. It was diagnosing the actual problem (generic content, not automation itself) and applying a targeted solution.

That's what this chapter teaches you. Not how to give up when AI doesn't work perfectly. How to diagnose what's wrong and fix it systematically.

The Diagnostic Framework: From Symptoms to Solutions

Most agents abandon AI implementations not because AI doesn't work, but because they don't know how to troubleshoot when problems arise.

The solution isn't guesswork. It's systematic diagnosis.

The Three-Question Diagnostic

When something isn't working, ask these questions in order:

Question 1: "What changed?"

Your Chapter 7 metrics tell this story:

- Which specific metric declined? (conversion rate? satisfaction scores? response time?)
- When did it start declining? (after implementing what?)
- What else changed at the same time? (new tool? different prompt? workflow adjustment?)

Question 2: "What's the actual problem?"

Separate symptoms from root causes:

- **Symptom:** Conversion rate dropped
- **Possible root causes:** Generic content, impersonal tone, response delays, technical errors, poor targeting

Question 3: "What's the smallest fix I can test?"

Don't rebuild everything. Identify one specific change to test:

- Refine one prompt
- Adjust one workflow step
- Add one personalization element
- Fix one integration issue

Test the fix. Measure results. Iterate if needed.

The Metrics-to-Diagnosis Connection

Your Chapter 7 metrics don't just measure success. They diagnose problems:

Metric Declining	Most Likely Root Cause	Where to Look First
Time Reclaimed	Workflow inefficiency or abandonment	Are you still using the automation? Is it slower than manual?
Conversion Rate	Content quality or personalization issues	Review AI outputs for generic language or a lack of specificity
Response Time	Technical failure or notification issues	Check automation triggers; test the notification system
Client Satisfaction	Over-automation or impersonal service	Audit communication frequency and the level of personalization
ROI	Mismatch between investment and value	Recalculate time savings and verify tool utilization

The principle: Your metrics tell you WHERE the problem is. This chapter tells you HOW to fix it.

 ⁄ **Future-Ready Checkpoint: Diagnostic Thinking for Agentic AI**

Common Obstacles: Real Estate-Specific Troubleshooting

Let's walk through the most common problems agents encounter and their proven solutions.

Category 1: Content Quality Issues

These problems show up when AI outputs don't meet your standards or client expectations.

Problem 1: Generic Listing Descriptions

Symptoms:

- Descriptions could apply to any property
- Lack of emotional appeal or unique character
- Broker or clients request revisions frequently
- Properties with great descriptions get more engagement; AI-generated ones get less

Diagnostic check (using metrics):

- Compare engagement rates (views, inquiries) for AI-written vs. manually written listings
- Track revision frequency (% of listings requiring significant rewrites)
- Monitor Days on Market (do AI-described properties sit longer?)

Root cause diagnosis:

Most often, this isn't an AI limitation. It's an input problem:

✗ **Poor prompt:** "Write a property description for a 3-bedroom house."

✅ **Effective prompt:**

> > Create a warm, inviting description for a 3-bedroom 1950s ranch in the Elmwood neighborhood of Berkeley, California.
>
> Unique features that set this property apart:
>
> - Original hardwood floors restored throughout
> - Chef's kitchen with vintage O'Keefe & Merritt stove
> - Built-in bookcases in living room
> - Private backyard with mature fruit trees and vegetable garden

- Walking distance to Ashby BART and Elmwood shopping district

Target buyers: Families or professionals who value mid-century character, walkability, and authentic charm over cookie-cutter modern.

Tone: Warm and authentic, celebrating character and craftsmanship.

Length: 250 words.

Ensure fair housing compliance throughout.

The Solution Framework:

1. **Provide 5+ truly unique features** (not "updated kitchen," but "chef's kitchen with vintage O'Keefe & Merritt stove")

2. **Include neighborhood specifics** (not "great location," but "walking distance to Ashby BART")

3. **Define target buyer psychology** (not "families," but "families who value character over cookie-cutter modern")

4. **Specify tone explicitly** ("warm and authentic" vs. "sophisticated and aspirational")

5. **Give examples of your voice** (include 2-3 paragraphs from your best past descriptions)

Story: Alex's Transformation

Alex's AI descriptions were consistently generic. His broker kept sending them back. He was about to give up on AI-assisted listing prep entirely.

His diagnosis: He'd been using the same basic template for all properties (luxury estates and starter condos got identical prompt structures).

His fix: Created three distinct prompt templates:

- **Luxury properties:** Emphasis on craftsmanship, exclusivity, lifestyle
- **Family homes:** Emphasis on community, schools, functional spaces

- **Investment properties:** Emphasis on cash flow, appreciation potential, tenant appeal

He also added a "voice bank" of his 10 best manual descriptions and instructed AI to match that style.

Result: Revision requests dropped from 60% to under 15%. Time per listing dropped from 45 minutes to 18 minutes. Quality improved because he was more thoughtful about what made each property unique.

Problem 2: Fair Housing Compliance Concerns

Symptoms:

- AI generates language that references demographics or protected classes
- Descriptions use coded language ("family neighborhood," "mature community")
- Concern about legal exposure from automated content

Diagnostic check:

- Audit 20-30 AI-generated property descriptions
- Highlight any language referencing or implying resident demographics
- Check for patterns across property types or neighborhoods

Root cause:

AI models learn from historical real estate content, which often contains biased or non-compliant language. If you don't explicitly instruct compliance, AI may reproduce problematic patterns.

The Solution:

1. Add explicit compliance instructions to every prompt:

> Critical compliance requirement: Ensure fair housing compliance throughout.

- Focus exclusively on property features, amenities, and accessibility

- Never reference or imply resident demographics, family status, or

---

protected classes

- Describe neighborhoods by proximity to services and amenities, not

resident profiles

- If in doubt, describe objective features only

---

2. Create a post-generation compliance checklist:

☐ Does the content describe only objective property features and amenities?

☐ Does it avoid references to "family neighborhood," "mature community," "young professionals," or similar coded language?

☐ Does it describe accessibility to services rather than resident profiles?

☐ Could any language be perceived as exclusionary or preferential?

☐ Have I removed any AI-generated content that feels questionable?

3. Use current AI models with better compliance training:

ChatGPT 5, Claude 4.5 Sonnet, and Gemini 3 have significantly improved fair housing awareness compared to earlier versions. Upgrade if you're using older models.

4. Implement human review for all marketing content:

Never publish AI-generated marketing materials without human review. This is both your professional responsibility and your legal protection.

Just-in-Time Learning: "AI Doesn't Remove Your Responsibility"

You are legally accountable for all content published under your name or license, regardless of whether AI helped create it. Review everything. If you wouldn't sign your name to it manually, don't publish it just because AI wrote it.

Problem 3: Inconsistent Brand Voice

Symptoms:

- Some communications sound like you, others don't
- Client feedback mentions "different tone" or "doesn't sound like you"
- Your marketing materials feel inconsistent across channels
- New AI outputs don't match your established brand

Diagnostic check:

- Review 10-15 AI-generated communications
- Compare to your best manual communications
- Identify specific language, tone, or style differences
- Check if different team members are using different prompts

Root cause:

Usually one of three issues:

1. You haven't clearly defined your voice for AI

2. Different prompts produce different voices

3. You're using multiple AI tools with different default styles

The Solution:

Create a comprehensive brand voice guide:

My Communication Style:

TONE: Professional yet conversational. Knowledgeable but never condescending. Enthusiastic without being pushy.

PERSONALITY: Trusted advisor who genuinely cares about client success.

Think: experienced friend giving honest guidance, not salesperson making a pitch.

LANGUAGE PREFERENCES:

- Use: "home" (not "property" or "unit")
- Use: "you'll love" (not "occupants will appreciate")

- Use: Simple, clear language (not industry jargon)
- Avoid: Corporate speak, buzzwords, hyperbole

EXAMPLES OF MY VOICE:

[Include 2-3 paragraphs from your best communications here]

When generating any content for me, match this voice precisely.

Save this as a reusable prompt preface. Paste it at the beginning of every AI interaction.

For teams: Store this in a shared document. Everyone uses the same voice guide. Consistency becomes automatic.

Test your fix:

Generate three different types of content (listing description, client email, social post) and verify they all sound consistent. If not, refine your voice guide until they do.

Category 2: Technical Integration Issues

These problems occur when tools don't work together smoothly or technical systems break down.

Problem 4: CRM Compatibility and Automation Failures

Symptoms:

- Duplicate lead entries in CRM
- Missing information from automated workflows
- Integrations randomly failing
- Data not syncing between platforms

Diagnostic check (using metrics):

- Review your "Response Time" metric (sudden increases often indicate automation failure)
- Check CRM for duplicate records (indication of integration issues)

- Test your workflows manually (do they execute reliably?)
- Review error logs if your automation platform provides them

Root cause:

Usually one of these technical issues:

1. API connection interrupted (credentials expired, permissions changed)
2. Field mapping errors (CRM fields don't match automation tool fields)
3. Data format incompatibility (dates, phone numbers formatted differently)
4. Automation platform limits reached (executions per month, API calls)

The Solution:

Step 1: Verify connections

- Log into your automation platform (Make, Zapier, etc.)
- Check connection status for all integrated tools
- Reconnect any showing errors or warnings
- Test each connection with a manual trigger

Step 2: Review field mapping

- Open your workflow
- Verify that data fields match exactly between systems
- Check for fields that may have been renamed in your CRM
- Update mappings if needed

Step 3: Implement error notifications

- Set up email alerts when workflows fail
- Most automation platforms support this (you should know immediately if something breaks)
- Test the notification system

Step 4: Build redundancy for critical workflows

- For high-priority automation (lead acknowledgment, transaction reminders), create backup manual processes

- If automation fails, you can temporarily handle manually until fixed
- Document the manual process clearly

Example: Charlotte's Integration Fix

Charlotte's lead automation suddenly stopped working. New leads weren't being added to her CRM. Her Response Time metric jumped from 12 minutes average to 8 hours (manual checking).

Her diagnosis: She'd updated her CRM password for security, which disconnected the API integration.

Her fix: Reconnected Make.com to her CRM with new credentials. Tested with sample lead. Working within 10 minutes.

Her prevention: Set up error notifications so she'd know immediately if it happened again, rather than discovering it days later through declining metrics.

Problem 5: Image Quality Degradation (Virtual Staging)

Symptoms:

- Virtual staging looks artificial or low-quality
- Furniture appears to float or have wrong perspective
- Colors are oversaturated or lighting is unnatural
- Clients or brokers reject the staged images

Diagnostic check:

- Compare original photos to staged versions side-by-side
- Identify specific quality issues (perspective, lighting, realism)
- Test with different source photos (is quality consistent or variable?)
- Check if recent tool updates changed output quality

Root cause:

Usually:

1. Low-resolution source photos (virtual staging AI needs high-quality inputs)

2. Poorly lit original photos (harsh shadows, dim rooms)

3. Extreme angles or wide-angle distortion in originals

4. Over-processing (running images through multiple AI tools)

The Solution:

Foundation: Start with professional-quality source photos

- Minimum 3000px width for staging
- Well-lit, properly exposed
- Straight angles (not extreme wide-angle distortion)
- Clean, empty rooms (easier for AI to stage)

Process:

1. Use professional photographer whenever possible

2. If shooting yourself, use proper lighting and avoid ultra-wide lenses

3. Process photos through ONE AI tool only (don't layer multiple enhancements)

4. Keep originals for comparison and disclosure

If quality is still poor:

- Try different virtual staging tools (ReimaginHome, Virtual Staging AI, Styldod)
- Some work better with certain property types or lighting conditions
- Most offer free trials; test with your typical photos before committing

Always disclose: Virtual staging must be clearly labeled in all marketing materials. This is both ethical practice and legal requirement in most jurisdictions.

Problem 6: Data Privacy and Security Concerns

Symptoms:

- Uncertainty about what data has been shared with AI tools
- Client concerns about confidentiality
- Potential compliance violations from uploading sensitive information
- Using free tools for client data

Diagnostic check:

- Audit what information you've uploaded to AI platforms in past 90 days
- Review AI tool privacy policies and data retention practices
- Identify any sensitive client information that may have been shared inappropriately
- Check whether you're using enterprise vs. consumer tool versions

Root cause:

Lack of clear data handling protocols. Most agents haven't defined rules for what data is safe to share with AI vs. what must remain private.

The Solution:

Establish a clear data security protocol:

✅ **Safe to share with professional AI tools:**

- Property addresses and public MLS information
- General market data and statistics
- Your own marketing materials and content
- Public property records and neighborhood information
- Anonymized client scenarios (no names, specific addresses, or identifying details)

❌ **Never share with AI tools:**

- Social Security numbers
- Financial statements or income information
- Bank account or credit card numbers
- Confidential client communications

- Legal documents with sensitive terms
- Information covered by confidentiality agreements
- Anything you wouldn't post publicly

Implementation:

1. Use paid, professional-grade AI tools for any business use (ChatGPT Plus, Claude Pro, Gemini Advanced)
2. Never use free consumer versions for client-related data
3. For highly sensitive work, use enterprise tools with data protection agreements
4. Remove or redact names and specific identifying details before processing documents
5. Maintain local encrypted backups of all original documents

Train yourself (and team): Create a simple one-page reference guide. "Can I put this in AI?" If uncertain, err on the side of caution.

Category 3: Client Experience Issues

These problems occur when automation affects how clients perceive your service.

Problem 7: Overly Automated Feel

Symptoms:

- Client feedback about feeling "processed" rather than served
- Comments like "this feels automated" or "are you even reading these?"
- Clients become less responsive over time
- Your satisfaction scores decline after implementing automation

Diagnostic check (using Chapter 7 metrics):

- Review Client Satisfaction scores before and after automation
- Track response rates to your communications (are clients engaging less?)
- Read client feedback comments for patterns

- Compare your automated communications to competitors' manual ones

Root cause:

Over-automation or insufficient personalization. You're automating content creation but forgetting to inject authentic human connection.

The Solution:

Implement the 60/40 Rule:

- 60% of communication volume can be AI-assisted (routine updates, follow-ups, marketing)
- 40% MUST be genuinely personal (key decisions, emotional moments, relationship building)

Add personalization checkpoints to automation:

Even automated communications should include:

- **Personal greeting referencing specific conversations:** "Based on what you mentioned about wanting a larger backyard..."
- **Specific property references:** Not "properties in your price range" but "the Riverside Drive property you loved"
- **Your authentic voice:** Use your brand voice guide, include personal observations
- **Strategic human touchpoints:** After 3 automated messages, schedule a personal call

Example fix for common automation:

✕ Generic automated follow-up:

"Hi [Name], just checking in on your home search. Have you found anything interesting? Let me know if you'd like to see more properties."

✅ Personalized automated follow-up:

"Hi [Name], I've been thinking about your comment that you loved the kitchen in the Cloverleaf Street property but wished it had a larger yard. I just found a new listing that has a nearly identical kitchen layout AND a 0.3-acre lot with mature trees. Want to see it this weekend? I think you'll love it."

The second feels personal because it references specific previous conversations and makes a specific relevant offer.

Story: Thomas's Recovery

Thomas automated his entire client communication system. Three months later, his satisfaction scores dropped from 9.1 to 7.8. Multiple clients said they "didn't feel like a priority."

His diagnosis: He'd automated too much. Even significant updates (offer accepted, inspection scheduled) were going out via templates without personal touches.

His fix:

- Reserved all major transaction milestones for personal calls (even if followed by automated summary email)
- Added weekly personal check-in calls for active clients (not automated)
- Modified automation to include specific references: "You mentioned being concerned about the roof. The inspection came back clean with an estimated 10 years remaining life."

Result: Satisfaction scores recovered to 9.3 within two months. Clients felt "even more connected" than before automation because his personal touches were more consistent and thoughtful.

Problem 8: Accuracy and Timeliness Issues

Symptoms:

- AI provides outdated market information
- Factual errors in generated content
- References to events or data that are no longer current
- Clients correct your information

Diagnostic check:

- Review recent AI outputs for factual accuracy
- Verify market data against current MLS sources
- Check if AI references outdated regulations or market conditions
- Track how often you need to correct AI-generated information

Root cause:

AI training data cutoff. Models don't know information after their training cutoff date. ChatGPT, Claude, and Gemini don't automatically know today's interest rates, recent sales, new developments, or current regulations.

The Solution:

Never rely on AI for current market data. Instead:

1. **Use AI for structure and format, you provide the data:**

✗ Poor prompt: "Create a market update for sellers in Austin."

✅ Effective prompt:

> > Create a market update for sellers in Austin.
>
> Use this current data I'm providing:
>
> - Current inventory: 4,230 active listings (up 15% from last year)
>
> - Average days on market: 38 days (up from 22 days last year)
>
> - Median sale price: $585,000 (up 3.2% year-over-year)
>
> - List-to-sale price ratio: 98.1% (down from 102.3% last year)
>
> Interpret these trends for sellers considering listing in the next 60-90 days.
>
> Tone: Factual but optimistic.
>
> Length: 400 words.

1. **Add verification checkpoints:**
   - Never publish AI-generated market data without verification
   - Cross-reference numbers against MLS, local board, or other authoritative sources
   - Include "last updated" dates on all market information
   - Set calendar reminders to refresh data quarterly
2. **Use current AI models:**
   - Newer models have more recent training data cutoffs

- ChatGPT 5 > ChatGPT 4
- Claude 4.5 Sonnet > Claude 3.5 Sonnet
- Gemini 3 > Gemini 2.5

3. **For real-time data, use appropriate tools:**
   - Use MLS for property data
   - Use Fred (Federal Reserve Economic Data) for economic indicators
   - Use local board reports for market statistics
   - AI is for analysis and presentation, not data sourcing

! Verification Checkpoint: Professional Responsibility

AI models don't automatically know today's interest rates, recent sales, new developments, or current regulations. Always verify time-sensitive information from current, authoritative sources. You are professionally responsible for accuracy, regardless of where the information originated.

Category 4: Prompt Engineering Issues

Many problems stem from unclear or ineffective prompts. Let's troubleshoot the most common patterns.

Problem 9: AI Output Doesn't Match Expectations

Symptoms:
- Results are consistently off-target
- Content is too long, too short, wrong tone, or wrong focus
- You're constantly re-prompting or manually rewriting

Diagnostic pattern:

Compare your prompt to these principles:

Is your prompt:
- ☐ Specific about desired outcome?

☐ Providing sufficient context?

☐ Including clear constraints (length, tone, format)?

☐ Defining target audience?

☐ Giving examples when possible?

The Solution: The Five-Element Prompt Structure

Every effective real estate prompt should include:

3. **Clear objective:**

   "Create a property description for..." / "Draft a follow-up email to..." / "Generate social media posts about..."

4. **Specific details:**

   Property features, client situation, neighborhood context, transaction stage

5. **Audience definition:**

   First-time buyers / luxury sellers / investors / relocating executives

6. **Tone and style:**

   Warm and conversational / sophisticated and aspirational / professional and direct

7. **Format constraints:**

   Length (word count), structure (paragraphs, bullets, headers), specific requirements

Before/After Example:

✖ **Vague prompt:**

"Write an email to a client about their offer."

☑ **Effective prompt:**

> Draft a follow-up email to buyer clients whose offer was just accepted.

Client context:

- First-time homebuyers (couple in their late 20s)
- Submitted offer yesterday at $525K on a $515K listing
- Competing offer situation (4 total offers)
- They're excited but also nervous about the process ahead

Message purpose:

- Congratulate them on offer acceptance
- Outline next steps in the transaction
- Reassure them about the process
- Set expectations for timeline

Tone: Warm and celebratory but also grounding and reassuring.

Length: 200-250 words.

Include: Specific next actions they need to take this week.

Match my communication style: [include example if available]

Result: First output is 80-90% ready to send with minimal personalization, vs. starting from scratch multiple times.

The Continuous Improvement Loop

Troubleshooting isn't just reactive (fixing problems as they occur). The best agents build continuous improvement into their practice.

The Monthly Refinement Ritual (30 minutes)

Week 1 of each month, review:

1. **Your Chapter 7 metrics (10 minutes):**
   - Are all metrics in the green zone?
   - Any declining trends over past 4 weeks?

- Any red flags requiring immediate attention?

2. **Your AI output quality (10 minutes):**

   - Review 5-10 recent AI-generated pieces
   - Rate quality on 1-10 scale
   - Identify patterns in what works vs. what doesn't
   - Note specific improvements to make

3. **Your prompt library (10 minutes):**

   - Which prompts produced best results this month?
   - Which needed significant editing?
   - Update your prompt templates based on learnings
   - Archive prompts that consistently underperform

Document improvements:

Keep a running log of refinements. Example:

March 2026 Improvements:

✓ Updated luxury listing prompt to emphasize lifestyle over features (revision rate dropped 40%)

✓ Added more specific neighborhood context to all property descriptions (client feedback improved)

✓ Refined follow-up email template to reference specific showing observations (response rate up 22%)

Next month focus:

- Improve social media engagement (currently underperforming)
- Test alternative virtual staging tool (quality concerns on some images)

The Quarterly Deep Dive (90 minutes)

Once per quarter:

1. **Comprehensive metrics review (30 minutes):**

   - Calculate quarterly averages for all five core metrics

- Compare to previous quarter and same quarter last year
- Identify biggest improvements and remaining gaps
- Set specific targets for next quarter

2. **Tool evaluation (30 minutes):**

- Are you using all tools in your stack effectively?
- Are there new tools that could improve results?
- Are there tools you're paying for but not using?
- Test one new tool or capability

3. **Process documentation (30 minutes):**

- Update your prompt library with quarterly improvements
- Document any workflow changes
- Share successful approaches with team (if applicable)
- Identify next area for automation expansion

The goal: Every quarter, your AI implementation should be measurably better than the previous quarter. Not just by accident, but by systematic refinement.

When AI Falls Short: Recognizing the Boundaries

Sometimes the problem isn't your implementation. It's that you're asking AI to do something it genuinely can't do well.

Recognizing AI's Limits in Real Estate

AI Limitation	Your Advantage	Best Practice
Knowledge cutoff dates (doesn't know recent events)	Real-time local market knowledge	Use AI for structure and analysis, but **YOU** supply current data

AI Limitation	Your Advantage	Best Practice
No true emotional intelligence or empathy	Reading client emotions, building trust, providing reassurance	Reserve emotionally significant communications for personal handling
Can't assess physical property condition from photos	Trained eye for quality, potential issues, neighborhood context	Use AI for visual enhancement, but **YOUR** assessment drives strategy
No understanding of relationship dynamics or negotiation psychology	Experience-based intuition, adaptive strategy, relationship leverage	Use AI for preparation and analysis; **YOU** conduct actual negotiations
Not current on local or recent regulatory changes	Professional responsibility, continuing education, broker guidance	Always verify AI suggestions against current regulatory requirements

The Recognition Framework: Five Tests

Before using AI for a task, run it through these tests:

1. **The Recency Test**

    "Does this require information that may have changed recently?"

    If YES → Verify data independently, don't rely on AI knowledge

2. **The Relationship Test**

    "Does this interaction significantly impact client trust?"

    If YES → Handle personally, don't automate

3. **The Judgment Test**

"Does this require professional experience and local market knowledge?"

If YES → Use AI for analysis prep, but YOU make the judgment call

4. **The Creativity Test**

"Does this require novel thinking beyond pattern recognition?"

If YES → Use AI for brainstorming options, but YOU develop creative strategy

5. **The Integrity Test**

"Could AI reliance here compromise my professional responsibilities?"

If YES → Handle personally, period

The Collaboration Model:

Task Type	AI's Role	Your Role
Information Gathering	Comprehensive data collection	Verify accuracy, add local insights
Content Creation	First drafts, formatting, consistency	Personal voice, strategic messaging, final approval
Data Analysis	Process large datasets, identify patterns	Interpret significance, apply local context
Client Communication	Templates, scheduling, routine follow-up	Key conversations, relationship building
Visual Content	Enhancement, organization	Selection, quality control, disclosure

Task Type	AI's Role	Your Role
Strategic Decisions	Option generation, pros/cons	Final judgment, client-specific adaptation

Remember: AI amplifies your expertise. It doesn't replace it. When in doubt, add more human judgment, not less.

Warning Signs of Over-Reliance

Your Chapter 7 metrics will often reveal over-reliance before you consciously recognize it. Watch for these patterns:

1. **Declining client face time despite increased "efficiency"**

   - Red flag: You're more "productive" but clients feel more distant
   - Solution: Schedule mandatory personal touchpoints regardless of automation

2. **Client feedback about generic or impersonal service**

   - Red flag: Satisfaction scores declining or stagnant
   - Solution: Audit all client-facing automation for personalization gaps

3. **Missing important details not captured in data**

   - Red flag: You used to notice things about properties/clients that you're not catching anymore
   - Solution: Reduce AI reliance for assessment tasks, increase personal observation

4. **Reduced confidence making judgment calls without AI**

   - Red flag: You feel uncertain about decisions you used to make instinctively
   - Solution: Deliberately make some decisions without AI input to maintain judgment skills

5. **Diminishing local market knowledge**

- Red flag: You're less aware of market shifts, new developments, neighborhood changes
- Solution: Schedule regular market tours, broker networking, community involvement

Action if you notice these signs:

- Increase personal client interactions immediately
- Reduce automation scope temporarily
- Reconnect with what made you successful before AI
- Remember: AI should amplify your strengths, not atrophy them

---

How Refinement Changed Everything

Let me share Nicole's continuous improvement journey over 12 months:

Month 1: Initial Implementation

- Set up basic AI workflows (listing descriptions, email templates, social content)
- Metrics: Time saved 6 hrs/week, conversion rate stable at 16%
- Assessment: "This is helpful but not revolutionary"

Month 3: First Refinement

- Problem identified: Social media engagement was declining (metrics don't lie)
- Diagnosis: AI-generated posts were too generic and infrequent
- Fix: Shifted from asking AI to "create content" to "here are 10 client testimonials and market insights; turn each into engaging social posts matching my voice"
- Result: Engagement up 47% over next 8 weeks

Month 6: Major Breakthrough

- Problem identified: Spending lots of time personalizing automated emails (defeating efficiency purpose)
- Diagnosis: Email templates were too generic
- Fix: Created detailed client profile system; emails automatically pull specific data (properties they've viewed, preferences they've stated, concerns they've raised) and reference them specifically

- Result: Personalization time dropped from 10 min to 2 min per email; response rates improved 31%

Month 9: Expansion

- Problem identified: Transaction coordination still consuming 8-10 hours per deal
- Diagnosis: Doing everything manually because automation seemed too complex
- Fix: Implemented Chapter 6 transaction workflow (timeline automation, milestone reminders, status updates)
- Result: Time per transaction dropped to 4-5 hours; client satisfaction scores increased (more consistent communication)

Month 12: Optimization

- Comprehensive review of entire year
- Total time reclaimed: 14 hours weekly (728 hours annually)
- Conversion rate: Up from 16% to 23%
- Client satisfaction: Up from 8.4 to 9.1
- Transactions closed: Up from 19 to 27 annually
- Work-life balance: "Dramatically improved. I actually take Sundays off now."

Nicole's key insight:

"Month 1, I thought I'd 'implemented AI.' But I'd really just dipped my toe in. The real transformation came from treating AI implementation as an ongoing practice, not a one-time project. Every month, I identified one thing that wasn't working optimally and fixed it. Twelve small improvements compounded into complete business transformation."

That's the continuous improvement mindset. Not perfection on Day 1. Systematic refinement over time.

What You've Accomplished

By working through this chapter, you've:

✓ **Learned diagnostic thinking** (symptom → root cause → targeted solution)

✓ **Built troubleshooting frameworks** for content quality, technical issues, and client experience

✓ **Mastered prompt engineering** principles that dramatically improve AI outputs

✓ **Understood AI's real limitations** and when human judgment is non-negotiable

✓ **Created continuous improvement systems** (monthly refinement, quarterly deep dives)

✓ **Recognized over-reliance warning signs** before they damage your business

✓ **Connected metrics to diagnosis** (using Chapter 7 data to identify problems early)

More importantly, you've developed **resilience**. You now know that problems with AI implementation aren't reasons to quit. They're opportunities to refine. Every agent encounters obstacles. Amplified agents troubleshoot them systematically and emerge stronger.

You're not just using AI. You're mastering it.

Transition to Chapter 9

You've built AI systems (Chapters 5-6). You're measuring their impact (Chapter 7). You can diagnose and fix problems as they arise (Chapter 8).

But here's the question that determines your long-term success:

What happens when AI evolves beyond what you've learned in this book?

When GPT-6 or GPT-7 launches with capabilities we can't imagine today? When agentic AI systems become mainstream and autonomous agents manage entire business functions? When the tools you've mastered are replaced by platforms that work completely differently?

Will you adapt seamlessly? Or will you need to relearn everything from scratch?

Chapter 9 answers this question. It teaches you:

- How to think about AI evolution (not just current tools)
- Which skills transfer regardless of technology changes
- How to prepare for agentic AI systems (2028-2030)
- What the amplified agent looks like in 2030
- How to future-proof your career in an AI-accelerated industry

You've learned to fish. Chapter 9 teaches you to recognize when the entire ocean is changing and how to adapt before most agents realize the shift is happening.

The most valuable skill isn't mastering today's AI. It's **learning how to keep mastering AI as it evolves.**

Let's build that skill.

Chapter 9: Future-Proofing Your Real Estate Career

The Agent Who Saw It Coming

In 2010, Karen was a top producer still using BlackBerry, fax machines, and physical lockboxes. She'd built her business on these tools. They worked. Why change?

Then smartphones disrupted everything. Her younger competitors were texting clients immediately, sending listings via mobile apps, and using digital lockboxes. Karen's clients started asking, "Why can't you just text me the listing?"

She resisted. "I'm not a tech person. I'm a relationship person. That's not how I work."

Her production dropped 40% in 18 months.

In 2013, facing reality, Karen hired a tech-savvy assistant to teach her. She learned smartphones. She adopted mobile apps. She rebuilt her systems around digital tools. Within two years, she'd recovered her production and then exceeded it.

In 2020, when COVID forced the entire industry online virtually overnight, Karen adapted smoothly. Virtual tours? She was ready. Digital closings? She'd been preparing. E-signatures? Already standard practice.

In 2023, after ChatGPT launched and AI became readily available, Karen didn't panic. She'd learned the pattern.

"Every 5-7 years, something fundamental changes," she told her mastermind group. "In the early 2010's was smartphones. 2020 was virtual everything. 2023 is AI. The agents who thrive aren't the ones who predict exactly what's coming. They're the ones who've built the capacity to adapt when change arrives."

By 2025, Karen was using AI extensively. Not because she could predict in 2010 that large language models would transform content creation. But because she'd learned **how to learn new technology** and integrate it strategically.

That's what this chapter teaches you. Not what AI will look like in 2030 (nobody knows for certain). But **how to think about technological evolution so you're ready for whatever emerges.**

The Fundamental Truth About Future-Proofing

Here's what we know for certain:

AI will evolve dramatically between now and 2030. Tools you're using today will be replaced. Capabilities that seem impossible now will become routine. Business models that work today will need adjustment.

We cannot predict specifics. Nobody in 2015 predicted that AI chatbots would be generating property descriptions and analyzing contracts by 2025. Specific predictions fail consistently.

But we can identify patterns. Technology evolution follows predictable patterns. Skills that matter evolve predictably. The agents who thrive during disruption share common mindsets and practices.

Your goal isn't clairvoyance. It's adaptability.

The Three-Layer Future-Proof Strategy

Layer 1: Transferable Skills (What Survives Every Technology Shift)

Core capabilities that remain valuable regardless of tools

Layer 2: Adaptive Mindset (How to Recognize and Respond to Change)

Mental frameworks for navigating technological evolution

Layer 3: Strategic Learning (How to Stay Current Without Drowning)

Systems for continuous improvement without overwhelm

Build these three layers, and you'll successfully navigate an AI-amplified future, regardless of which specific tools win or lose.

⟋ Future-Ready Checkpoint: The Meta-Skill

The most valuable skill isn't mastering ChatGPT 5 or Claude 4.5 or whatever emerges next. It's **learning how to quickly evaluate, adopt, and integrate new tools when they appear.** You've been practicing this throughout this book (assessing tools in Chapter 4, implementing workflows in Chapters 5-6, troubleshooting in Chapter 8). That meta-skill transfers to whatever comes next. Specific tools change. The evaluation and integration process doesn't.

Layer 1: Transferable Skills (What AI Will Never Replace)

Technology changes. Human nature doesn't. Certain capabilities will remain valuable in 2030 for the same reasons they were valuable in 1990.

The Unchanging Core: What Clients Actually Hire You For

The 2024 NAR data we examined in Chapter 7 reveals what clients value most when choosing agents:

Buyers prioritize:

- Agent's experience: 21%
- Honesty and trustworthiness: 19%
- Reputation: 15%

Sellers prioritize:

- Reputation of agent: 35%
- Agent is honest and trustworthy: 21%
- Agent's knowledge of the neighborhood: 10%

What buyers wanted help with most:

- Finding the right home: 49%
- Negotiating terms: 14%

- Price negotiations: 11%

Most valued benefits from their agent:

- Helping buyer understand the process: 61%
- Pointing out unnoticed features/faults: 58%
- Negotiating better contract terms: 46%

Notice what's NOT on these lists:

- "Responded within 2 minutes"
- "Generated great property descriptions"
- "Had the best technology"
- "Automated their follow-up"

Clients hire agents for judgment, trust, local knowledge, and negotiation skill. AI makes you more efficient at delivering those things. It doesn't replace them.

The Five Skills That Compound Over Time

These capabilities become MORE valuable as AI handles routine tasks:

1. Judgment in Ambiguous Situations

What it is: The ability to make sound decisions when data is incomplete, conflicting, or insufficient.

Why AI can't replicate it: AI analyzes patterns in data. Novel situations without clear patterns require human intuition built from experience.

Examples in real estate:

- Pricing a truly unique property with no good comparables
- Assessing whether a buyer is genuinely qualified vs. just pre-approved on paper
- Evaluating whether a transaction risk is worth taking
- Deciding when to push hard in negotiation vs. when to compromise
- Recognizing when a client needs to hear hard truth vs. encouragement

How to develop it:

- Deliberately reflect on past decisions (what worked? what didn't? why?)
- Seek feedback from experienced agents on your judgment calls
- Study case examples where judgment was critical
- Practice explaining your reasoning (articulating "why" improves judgment)
- Develop pattern recognition through volume of transactions

The AI relationship:

AI provides data and analysis. You apply judgment to interpret what it means in context and decide what action to take.

2. Trust Building Through Authentic Presence

What it is: The ability to create genuine human connection that makes clients feel safe during major life decisions.

Why AI can't replicate it: Trust isn't created by efficiency. It's created by demonstrated care, consistent integrity, and authentic human interaction.

Examples in real estate:

- The conversation where a client admits they're scared about the financial commitment
- The moment when a seller breaks down about leaving their family home
- The negotiation where both parties need someone they trust to guide them
- The call at 9 PM when a client has cold feet and needs reassurance
- The relationship that generates referrals 5 years after a transaction

How to develop it:

- Practice active listening (truly hearing, not just waiting to respond)
- Follow through obsessively on commitments (trust is built through consistency)

- Demonstrate vulnerability and authenticity (perfection creates distance)
- Show up for clients beyond transactions (relationship, not transaction focus)
- Prioritize long-term reputation over short-term gain (integrity compounds)

The AI relationship:

AI handles routine communications efficiently. You focus your energy on trust-building moments that matter most.

3. Local Market Knowledge Beyond Data

What it is: Deep understanding of neighborhood dynamics, future development, community culture, and market psychology that can't be captured in databases.

Why AI can't replicate it: True local expertise comes from being embedded in a community over time. AI can analyze public data, but not feel a neighborhood's vibe or predict how character is changing.

Examples in real estate:

- Knowing which streets flood in heavy rain despite not being in official flood zones
- Understanding school district boundary politics and likely future changes
- Recognizing which neighborhoods are gentrifying vs. which are stable
- Identifying micro-markets (one side of the street differs dramatically from the other)
- Predicting how planned development will impact property values and desirability

How to develop it:

- Spend time physically in your market (regular neighborhood tours, not just showings)
- Attend city planning meetings and community forums
- Build relationships with longtime residents, business owners, city officials

- Document patterns you notice over years (your proprietary knowledge base)
- Study historical development patterns in your area

The AI relationship:

AI organizes and analyzes public market data. You layer on the impossible-to-quantify local insights that create differential value.

4. Strategic Creativity Under Constraints

What it is: The ability to generate novel solutions to unique problems, especially when standard approaches don't work.

Why AI can't replicate it: AI recognizes patterns in existing data. True creativity requires combining concepts in ways that don't yet exist in the training data.

Examples in real estate:

- Structuring a deal that works for both parties when conventional approaches fail
- Marketing an unusual property that doesn't fit standard templates
- Solving title issues or transaction obstacles that threaten deals
- Creating win-win solutions when buyer and seller seem impossibly far apart
- Developing unique positioning for yourself in a crowded market

How to develop it:

- Study solutions from other industries (how would a different field solve this problem?)
- Practice constraint-based thinking (what if I couldn't use the obvious solution?)
- Build a "creative solutions" library from your experiences and others'
- Collaborate with creative thinkers from different backgrounds
- Give yourself permission to consider unconventional approaches

The AI relationship:

AI generates options based on patterns. You select, combine, and innovate beyond what AI suggests to create truly novel solutions.

5. Emotional Intelligence in High-Stakes Interactions

What it is: The ability to read emotional states, navigate conflict, regulate your own emotions, and guide others through stressful decisions.

Why AI can't replicate it: Buying and selling homes is deeply emotional. AI can simulate empathy in text, but can't authentically connect with human fear, joy, anxiety, and stress.

Examples in real estate:

- Recognizing when a client's objection isn't about the issue they're stating
- De-escalating conflict between buyer and seller during tense negotiations
- Supporting clients through transaction stress without absorbing their anxiety
- Knowing when to be encouraging vs. when to deliver difficult reality
- Building consensus among multiple decision-makers with competing priorities

How to develop it:

- Study emotional intelligence frameworks (read books like "Emotional Intelligence 2.0")
- Practice labeling emotions (yours and others') to build awareness
- Seek feedback on how others experience your emotional presence
- Reflect on interactions where emotions ran high (what worked? what didn't?)
- Develop stress management practices so you can stay regulated when clients aren't

The AI relationship:

AI handles transactional communication. You handle emotionally significant interactions where human connection is non-negotiable.

The Amplified Agent's Advantage

Notice the pattern: AI doesn't diminish these skills. It amplifies them.

Without AI: Your time is fragmented across routine tasks and high-value activities. You have less energy for judgment, relationship building, creative problem-solving, and emotional intelligence because you're exhausted from administrative work.

With AI: Routine tasks are handled efficiently. You have more mental bandwidth for the activities that differentiate you and build long-term value.

The agents who master AI while developing these five core skills become exponentially more valuable than agents who only have one or the other.

That's the amplified agent advantage.

Layer 2: Adaptive Mindset (How to Navigate Change)

Specific tools and techniques will evolve. Your mindset about that evolution determines whether you thrive or struggle.

The Four Mindset Shifts

Shift 1: From "Mastery" to "Fluency"

Old mindset: "I need to completely master each tool before moving forward."

New mindset: "I need functional fluency with tools, then continuous refinement through use."

Why it matters: Technology evolves too quickly for traditional mastery. By the time you've "mastered" one tool, three better ones have emerged. Fluency means you can use a tool effectively for your needs while remaining open to evolution.

In practice:

- Aim for 80% competency, not 100% perfection
- Learn through doing, not just studying
- Accept that you'll discover better approaches as you go
- Focus on problems solved, not features memorized

Example: You don't need to know every ChatGPT capability. You need to know how to get good property descriptions, client emails, and market analysis. That's fluency. Mastery can wait.

Shift 2: From "This Tool" to "This Capability"

Old mindset: "I use ChatGPT" or "I use Claude."

New mindset: "I use AI language models for content generation and analysis" (currently via ChatGPT, but I could switch).

Why it matters: Tools come and go. Capabilities persist. By thinking in terms of capabilities, you become tool-agnostic and can pivot smoothly when better options emerge.

In practice:

- Define what you need accomplished (capability), not which tool to use
- Evaluate tools based on how well they deliver needed capabilities
- Remain willing to switch tools when a better option proves itself
- Build workflows around capabilities, with tools as interchangeable components

Example: Your workflow requires "AI-generated, personalized follow-up emails." Today that's ChatGPT. In three years, it might be something else. Your workflow design doesn't need to change, just the tool filling that capability role.

Shift 3: From "Learn It All Now" to "Learn What's Next"

Old mindset: "I need to learn everything about AI immediately."

New mindset: "I need to learn what serves my business goals now, with a system for learning what comes next."

Why it matters: You'll never "finish" learning AI. It's continuously evolving. Accepting this frees you from trying to learn everything and allows you to focus strategically.

In practice:

- Identify your current highest-leverage opportunity (what would help most right now?)
- Implement deeply before moving to the next capability
- Build learning into your routine (Chapter 8 continuous improvement rituals)
- Stay aware of developments without feeling pressure to adopt everything immediately

Example: Maybe automated listing descriptions are your current focus. Become excellent at that before trying to build complex agentic workflows. Sequential mastery beats simultaneous mediocrity.

Shift 4: From "Technology vs. Humanity" to "Technology Amplifying Humanity"

Old mindset: "AI is either going to replace me or I need to resist it to protect my humanity."

New mindset: "AI handles routine tasks so I can focus more energy on distinctly human value creation."

Why it matters: The adversarial framing creates resistance. The collaborative framing creates opportunity. AI isn't the enemy of human-centered real estate practice. It's the tool that enables it.

In practice:

- Frame AI as your assistant, not your replacement
- Use efficiency gains to increase quality, not just quantity
- Measure success by client experience and life margin, not just time saved
- Position yourself as tech-enabled human expert, not tech-dependent automaton

Example: "I use AI to handle research and drafting so I can spend more face time with clients understanding their actual needs." That's amplification, not replacement.

The Pattern Recognition Framework

Technology disruptions follow predictable patterns. Recognizing them helps you adapt proactively.

Pattern 1: Initial Hype Exceeds Reality

When new technology launches, capabilities are overstated. Early adopters claim revolution. Then reality settles in.

Your response: Explore new technology with curiosity but implement with caution. Test small. Validate value. Scale when proven.

Pattern 2: Rapid Capability Improvement

Early versions of new technology are clunky. Within 12-24 months, capabilities improve dramatically.

Your response: If a tool almost solves your problem but not quite, revisit it in 6-12 months. Patience often beats premature abandonment.

Pattern 3: Commoditization of Yesterday's Innovation

What's expensive and cutting-edge today becomes cheap and standard tomorrow.

Your response: Don't over-invest in tools at premium prices unless they deliver immediate ROI. Wait for price drops if you can.

Pattern 4: Integration Beats Point Solutions

Early in a technology cycle, specialized point solutions dominate. Over time, integrated platforms absorb capabilities.

Your response: Build workflows around capability categories, not individual tools. Makes transitions easier as integration happens.

Pattern 5: Human Backlash to Over-Automation

When automation becomes excessive, humans push back and demand authentic human connection.

Your response: Never automate so much that clients feel processed. Maintain strategic human touchpoints regardless of efficiency pressure.

The Adaptation Cycle

When significant new capabilities emerge (as they will regularly between now and 2030), use this cycle:

Phase 1: Awareness (Week 1)

- Notice the new capability through your information sources
- Understand at high level what it does and why it might matter
- Assess: relevant to real estate? relevant to my practice specifically?

Phase 2: Evaluation (Weeks 2-3)

- Test the tool with non-critical tasks
- Compare to current approaches
- Assess: does this solve a real problem? is ROI positive?

Phase 3: Pilot (Weeks 4-6)

- Implement in limited context alongside existing processes
- Measure actual impact using your Chapter 7 metrics
- Decide: adopt fully, adopt partially, or abandon?

Phase 4: Integration (Months 3-6)

- If adopting, build into standard workflows
- Train (yourself and team if applicable)
- Document and refine
- Optimize based on Chapter 8 continuous improvement

Phase 5: Maintenance (Ongoing)

- Monitor performance

- Stay aware of competitive alternatives
- Reassess periodically (is this still the best option?)

This cycle repeats continuously. By 2030, you'll have gone through it 15-20 times with different tools and capabilities. Each time, you'll get faster and better at it.

That's how adaptability becomes a competitive advantage.

Layer 3: Strategic Learning (Staying Current Without Drowning)

Between now and 2030, you could spend 40 hours weekly reading about AI developments. Or you could spend 2 hours monthly and get 90% of the value.

The Minimal Viable Information Diet

Monthly: 60 Minutes Total

Week 1 (20 minutes): Industry Pulse Check

- Scan Inman Technology section headlines (5 min)
- Check one AI-focused newsletter like "The Rundown AI" (10 min)
- Note anything directly relevant to real estate (5 min)

Week 2 (15 minutes): Tool Discovery

- Browse Product Hunt or G2 for new real estate-relevant AI tools (10 min)
- Add interesting finds to your evaluation list (5 min)

Week 3 (15 minutes): Peer Learning

- Check one online community for agent discussions about AI (10 min)
- Contribute one insight or question (5 min)

Week 4 (10 minutes): Reflection

- Review your notes from the month

- Identify one thing to test in the next 30 days
- Update your technology tracking document

Quarterly: 2-3 Hours

Tool Testing (90 minutes)

- Pick one new tool from your evaluation list
- Test with actual work tasks
- Document results in your Chapter 8 tracking system
- Decide: adopt, revisit later, or abandon

Deep Dive Learning (60 minutes)

- Take one focused tutorial or watch an in-depth implementation video
- Could be about a tool you're already using (going deeper) or a new capability

Strategic Review (30 minutes)

- Review your Chapter 7 metrics quarterly trends
- Assess whether your current tool stack is serving you well
- Plan any changes for next quarter

Annually: 4-6 Hours

Comprehensive Tech Stack Audit (3 hours)

- Review every tool you're paying for
- Eliminate anything not delivering clear ROI
- Research whether better alternatives exist for key capabilities
- Update your technology strategy for the year ahead

Learning Investment (2-3 hours)

- Attend one webinar, virtual conference session, or in-person event
- Focus on emerging trends and capabilities that might matter in next 12-24 months
- Network with other forward-thinking agents

Budget Review (1 hour)

- Calculate actual ROI on AI investments
- Reallocate budget based on what's working vs. not

- Plan investment for new capabilities in year ahead

Total time investment: 16-20 hours annually. That's 0.4% of a 2,000-hour work year. Achievable for even the busiest agent.

The Signal vs. Noise Filter

How do you know what information actually matters vs. what's just hype?

Ask these three questions about any new development:

Question 1: "Is this solving a real problem I have?"

- If NO: Interesting but not actionable. File for future awareness.
- If YES: Worth deeper evaluation.

Question 2: "Is this ready for practical use today, or still experimental?"

- If EXPERIMENTAL: Track it, but wait for maturity.
- If READY: Proceed to testing phase.

Question 3: "Would this deliver measurable improvement over current approach?"

- If NO or UNCLEAR: Not worth the switching cost.
- If YES: Pilot it using your Chapter 8 framework.

Most developments fail at least one of these questions. That's fine. Your job isn't to adopt everything. It's to selectively adopt what meaningfully improves your practice.

The Peer Learning Multiplier

You don't need to discover everything yourself. Build a small network of tech-forward agents who share discoveries.

The 5-Person Learning Network:

Find 4-5 agents (not in your direct market to avoid competition concerns) who are:

- At similar business stage and sophistication
- Genuinely curious about AI and technology
- Willing to test things and share learnings

- Different enough to bring diverse perspectives

Structure:

- Monthly 60-minute video call
- Each person shares: one win, one challenge, one discovery
- Group troubleshoots challenges collaboratively
- Shared document tracking what everyone is testing

Value multiplier: 5 people each testing one new tool monthly = you benefit from 5x the experimentation at 1x the time cost.

What "Agentic AI" Probably Means for Real Estate

We can't predict specific tools. But we can identify probable capability evolution based on current trajectories.

What's Likely to Emerge

More autonomous task completion:

Instead of "write me a property description," you'll likely say "prepare this listing for MLS" and an AI system will generate description, select photos, write copy, suggest price range, and queue everything for your review.

Better context retention:

AI systems will remember your past interactions, client preferences, and business patterns without you needing to re-explain context constantly.

Multi-step workflow execution:

You'll design complex workflows (like those in Chapter 6) more easily, and AI will execute them with less manual supervision.

Proactive suggestions:

Rather than waiting for you to ask, AI will likely suggest actions: "Three of your past clients own homes in an area showing strong appreciation. Consider reaching out about potential sales."

Voice interaction:

Expect AI assistants you can talk to naturally, not just type at. Could transform how you interact with technology while driving between showings.

What This Means for You

Your role elevates further: More strategic oversight, less manual execution. You become the architect and quality controller. AI becomes the execution engine.

Learning curve gets easier: As AI becomes more intuitive and conversational, technical skill requirements decrease. Implementation gets more accessible.

Competitive differentiation shifts: Everyone will have access to powerful AI. Your differentiation comes from how you design systems and apply judgment, not from having AI access.

Ethics and oversight become critical: More powerful automation means more potential for problems at scale. Your responsibility for monitoring and controlling AI outputs intensifies.

How to Prepare

1. **Master workflow design thinking (Chapter 6 skillset)**

    The ability to think systematically about how work should flow will matter even more when AI can execute complex processes.

2. **Develop strong quality standards (Chapter 8 refinement discipline)**

    Autonomous systems require clear quality criteria. Your taste and standards become your primary control mechanism.

3. **Build robust metrics systems (Chapter 7 discipline)**

    You'll supervise AI by monitoring outcomes, not watching processes. Metrics become your primary feedback mechanism.

4. **Strengthen core human skills (Layer 1 of this chapter)**

As AI handles more execution, your differentiation comes entirely from judgment, relationships, creativity, and emotional intelligence.

5. **Stay ethically grounded (Chapter 1 principles)**

    More powerful tools create more potential for harm. Your ethical foundation matters more, not less.

The pattern: Skills you're building now are exactly what you'll need in 3-5 years from now, just applied to more powerful technology.

The Mindset That Survives Everything

Let me share the most important insight from Karen's story at the beginning of this chapter.

She didn't thrive because she predicted smartphones in 2010 or AI in 2020. She thrived because she learned one meta-lesson:

"The technology will change. My willingness to learn must not."

The Three Commitments

If you commit to these three practices, you'll navigate whatever emerges between now and 2030:

Commitment 1: Continuous Learning

Not binge-learning when disruption forces you. Regular, sustainable learning as ongoing practice.

"I dedicate 60 minutes monthly to staying current with technology relevant to my practice."

Commitment 2: Thoughtful Implementation

Not adopting every shiny new tool. Not resisting all change. Discerning evaluation of what actually serves your clients and business.

"I assess new tools against my business goals, test them systematically, and implement what delivers clear value."

Commitment 3: Human-Centered Focus

Not letting efficiency become the only goal. Using technology to enhance human connection, not replace it.

"I use AI to handle routine work so I can focus more energy on judgment, relationships, and the irreplaceable human elements of this profession."

Those three commitments matter more than any specific tool recommendation in this book.

The Long View

Real estate has survived:

- The internet (agents worried about disintermediation in 1995)
- Zillow and automated valuations (agents worried about obsolescence in 2006)
- Virtual tours and remote closings (agents worried about relevance in 2020)
- AI and automation (agents worrying about replacement today)

Agents didn't disappear. The role evolved.

The agents who evolved with each shift thrived. The ones who resisted struggled. But the profession adapted and survived because **human judgment, local expertise, relationship building, and negotiation skill remain valuable.**

AI makes those capabilities more valuable, not less, by removing the noise that diluted them.

What You've Accomplished

By working through this chapter, you've:

✓ **Identified transferable skills** that will remain valuable regardless of technological change

✓ **Developed adaptive mindset** for navigating evolution without panic

✓ **Created sustainable learning system** that keeps you current without overwhelming you

✓ **Understood probable AI evolution** without depending on specific predictions

✓ **Built the meta-skill** of learning how to learn new technology quickly

✓ **Established three commitments** that will carry you through 2030 and beyond

More importantly, you've shifted from **"what will the future look like?"** to **"how do I build the capacity to thrive regardless of what emerges?"**

That's the only real future-proofing.

Transition to Chapter 10

You've learned the concepts (Chapters 1-2). You've integrated AI into daily workflow (Chapters 3-4). You've mastered specific applications (Chapter 5). You've built advanced systems (Chapter 6). You've measured impact (Chapter 7). You've troubleshot problems (Chapter 8). You've prepared for evolution (Chapter 9).

One question remains:

What should you do tomorrow?

Knowledge without implementation is just entertainment. Chapter 10 transforms everything you've learned into a concrete 90-day action plan.

You'll get:

- Week-by-week implementation roadmap (customized to your starting point)
- Decision trees for choosing what to implement first
- Accountability systems that ensure follow-through
- Integration guidance for teams and brokerages
- Specific milestones and success criteria

You know WHAT to do and WHY it matters. Chapter 10 tells you exactly HOW to start and WHEN to progress to each next level.

From knowledge to action. From reading to results. From understanding amplified agents to becoming one.

Let's build your implementation plan.

Chapter 10: Your Action Plan

The Conversation You Need to Have With Yourself

You've just absorbed 9 chapters covering AI foundations, daily integration, tool selection, specific applications, advanced workflows, metrics, troubleshooting, and future-proofing. That's a lot.

Maybe you're energized. Maybe you're overwhelmed. Maybe you're somewhere in between.

Before you dive into implementation plans and checklists, let's have an honest conversation about what happens next.

Here's what typically happens after agents read books like this:

Scenario 1: The Enthusiastic Sprint

You're fired up. You're going to implement everything. You spend a weekend setting up 7 new tools, creating 15 prompts, building automation workflows. By Tuesday, you're exhausted and back to your old way of working. The tools sit unused. You feel like you failed.

Scenario 2: The Paralyzed Perfectionist

You want to do this right. So, you re-read chapters. You research additional tools. You create elaborate plans. Weeks pass. You haven't implemented anything because you're still trying to figure out the "perfect" approach. Analysis paralysis wins.

Scenario 3: The Skeptical Abandonment

You try one thing. It doesn't work perfectly on the first attempt. "See? This AI stuff isn't for me." You close the book and never return to it.

Scenario 4: The Sustainable Builder

You choose ONE thing that would genuinely help your business right now. You implement it simply. You give it two weeks to show value.

Then you refine it or try something else. Six months later, you've integrated AI meaningfully into your practice without burning out.

This chapter is designed to help you be Scenario 4.

Not because you need to follow my exact prescription. But because **sustainable change beats perfect plans that never happen.**

The Right AMMA Path For You: Assess > Map > Monitor > Adjust

The following will guide you through an assessment of your what you want AI to do for you, why, and how much you are willing to invest in that journey (time, attention and cost). From those results, you will choose a path to take in expanding your AI proficiency and use that will equate to a roadmap option designed to help get you there. Lastly, you will find guidance on how to assure continued progress along the way and to tweak what is and isn't working.

Progress through this action plan on your own or use the URL or QR code below to benefit from our personalized, online interactive version enhanced with up-to-date information.

https://go.howdoiai.pro/re-action-plan

Start With Why (Your Why, Not Mine)

Before touching any tool or creating any prompt, answer three questions for yourself:

Question 1: What problem am I trying to solve?

Not "I should use AI because everyone says I should." That's not sustainable motivation.

What specific pain point in your current business would you pay to eliminate?

Common answers:

- "I spend 6 hours weekly on listing prep and it's eating my life"
- "I'm terrible at consistent follow-up and I'm losing deals because of it"
- "I can't keep up with social media, and my online presence is suffering"
- "Transaction coordination is my nightmare; things slip through the cracks"
- "I want to serve more clients without working 60-hour weeks"

Your answer: _____

Why this matters: Clear problems drive focused solutions. Vague goals ("be more efficient") lead to scattered implementation and abandonment.

Question 2: What's my real constraint right now?

Is it time? "I'm maxed out. I literally cannot add more complexity to my life right now."

Is it money? "I need to see ROI before investing in premium tools."

Is it confidence? "I'm not tech-savvy. I'm afraid I'll waste time on things I can't figure out."

Is it clarity? "I don't know where to start. Too many options overwhelm me."

Is it energy? "I'm exhausted. I need simple wins, not complex projects."

Your constraint: _____

Why this matters: Your constraint determines your starting point. There's no universal "right" answer. An agent with unlimited time but no budget starts differently than an agent with budget but no time.

Question 3: What does success look like for me?

Not what should success look like. Not what success looks like for the agent in Chapter 5's case study. For YOU.

For some agents, success is:

- "I leave the office by 5 PM to have dinner with my family"
- "I double my transaction volume without hiring staff"
- "My client satisfaction scores hit 9+ consistently"
- "I stop dreading administrative work"
- "I feel confident I'm not falling behind competitors"

Your success vision: _____

Why this matters: Success isn't one-size-fits-all. Life margin matters to one agent. Revenue growth matters to another. Client experience matters to a third. All are valid. But YOUR definition guides YOUR choices.

The Self-Assessment: Where Are You Really?

Let's figure out your actual starting point. Be honest. Nobody's grading this.

Current AI Usage Level

Level 0: Complete Beginner

- I've never used ChatGPT or any AI tool for work
- The terminology in this book was mostly new to me
- I need to start with absolute basics

Level 1: Curious Experimenter

- I've tried ChatGPT a few times but don't use it regularly

- I understand the concepts but haven't built any workflows
- I need help with consistent implementation

Level 2: Active User

- I use AI tools weekly for specific tasks
- I have some workflows but they're not optimized
- I'm ready to level up my sophistication

Level 3: Power User

- AI is integrated throughout my business
- I came to this book to find advanced techniques and refinements
- I'm ready for complex automation and workflow design

Your level: _____

Available Time Reality Check

How much time can you ACTUALLY dedicate to AI implementation in the next 30 days?

Be realistic. You have listings, showings, closings, family obligations, and life happening.

Option A: "I'm slammed. Maybe 2-3 hours total this month."

→ Your path: One simple, high-impact implementation. That's it.

Option B: "I could carve out 1 hour per week."

→ Your path: One focused implementation per week.

Option C: "I'm between busy seasons. I can invest 10-15 hours this month."

→ Your path: Comprehensive setup of multiple workflows.

Option D: "I'm making this a priority. 20+ hours available."

→ Your path: Full integration including advanced automation.

Your reality: _____

There's no shame in Option A. Progress beats perfection. One meaningful change implemented beats ten perfect plans that never happen.

Budget Reality Check

What can you invest in AI tools in the next 90 days?

Option A: "$0-25/month. I need to prove value with free tools first."

> → Your path: Free tier of ChatGPT, free tools, minimal investment

Option B: "$25-100/month. I can do basic subscriptions."

> → Your path: ChatGPT Plus or Claude Pro, Canva, maybe one specialized tool

Option C: "$100-300/month. I'm willing to invest if ROI is clear."

> → Your path: Premium LLM, design tools, automation platform, maybe CRM upgrades

Option D: "$300+/month. I'll invest in comprehensive stack."

> → Your path: Full toolkit including specialized real estate AI platforms

Your budget: _____

Remember: Chapter 7's ROI calculations showed most agents get 500-1000%+ ROI on even modest investments within 6 months. But you need to start where you are, not where you wish you were.

The Resonance Test

Which chapters or sections genuinely resonated with you? Which made you think, "YES, that's exactly what I need"?

Go back through the book. Mark the sections that spoke to you.

Maybe it was:

- Chapter 5's property marketing section (writing better listing descriptions, faster)
- Chapter 6's client journey automation (never dropping leads again)
- Chapter 7's time reclamation calculations (getting my life back)
- Chapter 8's prompt engineering (making AI sound like me)

Your top 3 resonant sections:

1.

2.

3.

Start there. Not where I tell you to start. Not where the "ideal" agent starts. Where YOU felt energy and possibility.

Your Personalized Path's Starting Point

Based on your responses to the varied self-assessment introspective, choose the path to begin:

Path 1: Complete Beginner, Minimal Time/Budget

Your reality: Never used AI, have maybe 2-3 hours available, and prefer a minimal budget.

Goal: Answer the question "Is AI actually useful for my real estate business?"

Week 1: Getting Started (30 minutes):

- Sign up for free ChatGPT
- Instead of using Google or other search engine for any real estate query, use ChatGPT

Week 2: First Real Use (60 minutes)

- Use Chapter 5's property description prompt template for your next listing
- Compare time spent: manual vs. AI-assisted

- Assess quality: good enough? needs editing?

Week 3: Expand Testing (45 minutes)

- Use Chapter 5's client email templates for three real communications
- Track response rates vs. your normal emails
- Note: does this feel helpful or not?

Week 4: Decision Point (30 minutes)

- Review your experience
- Calculate time saved (even if small)
- Decide: continue and expand? Or this isn't for me?

Success criteria: You've tested AI in real business situations and have data (time, quality, client response) to decide next steps.

Month 2 decision:

- If helpful → Upgrade to ChatGPT Plus ($20/month) and expand to social media content
- If not helpful → Revisit why (was it the prompts? the approach? the tool?) before abandoning

Why this works: You're not trying to transform your entire business. You're answering one question: "Can this actually help me?" Small stakes. Clear test. Tangible value or not.

Path 2: Curious Experimenter, 1 Hour Weekly, Modest Budget

Your reality: You've tried AI but don't use it consistently, have 4-5 hours available this month, and willing to invest $25-100 per month in AI tools.

Goal: Integrate AI into daily workflow consistently

Your starting point (Month 1):

Week 1 (60 min setup):

- Subscribe to ChatGPT Plus or Claude Pro ($20-25/month)
- Set up your brand voice guide (Chapter 8 template)

- Create your first three saved prompts (property description, client email, market update)
- Establish baseline metrics (Chapter 7): time currently spent on content creation weekly

Week 2 (60 min implementation):

- Use AI for all listing content this week (descriptions, social posts, email announcements)
- Track time spent vs. your usual manual approach
- Note quality issues for refinement

Week 3 (60 min refinement):

- Review what worked and what didn't
- Refine your prompts based on Chapter 8 guidance
- Add Canva for visual content (free tier to start)
- Create one social media post series using AI + Canva

Week 4 (60 min expansion):

- Implement one follow-up email automation (Chapter 5 templates)
- Set up basic tracking (Chapter 7's Weekly Scorecard)
- Plan Month 2 priorities

Success criteria: AI is now part of your regular workflow. You're seeing measurable time savings. You have systems for ongoing use.

Month 2 decision:

- Expand to transaction coordination automation (Chapter 6)
- Add virtual staging tool if property marketing is your focus
- Integrate with CRM if you have one

Why this works: Focused weekly implementation builds momentum. Each week delivers immediate value while setting up the next step. You're building capability progressively, not trying to do everything at once.

Path 3: Active User, 10-15 Hours Available, $100-300 Budget

Your reality: Already using AI for basic tasks, ready to level up, and can invest real time and budget.

Goal: Build sophisticated workflows that multiply impact

Your starting point (Month 1):

Week 1 (3 hours):

- Audit current AI usage (what's working? what isn't? what's missing?)
- Complete Chapter 7's baseline metrics for your five core KPIs
- Upgrade tools as needed (premium LLM, design platform, maybe automation tool)
- Create comprehensive prompt library for all regular content needs

Week 2 (3 hours):

- Design workflow architecture for biggest pain point
- Implement one major workflow automation (choose from Chapter 6)
- Lead qualification automation, OR
- Transaction timeline automation, OR
- Post-closing relationship sequence
- Test thoroughly before going live
- Create quality control processes

Week 3 (3 hours):

- Implement second workflow or optimize first
- Expand content creation to all channels (property, email, social, market updates)
- Implement quality control checklist (Chapter 8)
- Train on one advanced technique (maybe GTM engineering basics from Chapter 6)

Week 4 (3-4 hours):

- Review Month 1 metrics
- Troubleshoot any issues using Chapter 8 frameworks
- Calculate ROI using Chapter 7 frameworks
- Document successful workflows

- Plan Month 2 advanced implementations

Success criteria: You've implemented at least one sophisticated workflow that operates reliably. Metrics show clear business impact. You're ready for advanced applications.

Month 2-3 focus:

- Build advanced multi-step workflows
- Implement GTM engineering if ready (Chapter 6 investor outreach example)
- Create comprehensive client journey automation
- Maybe begin team training if applicable

Why this works: You have the time and budget to implement meaningfully. You're not dabbling. You're systematically building an AI-powered business infrastructure that compounds over time.

Path 4: Power User, 20+ Hours, $300+ Budget

Your reality: You're already utilized AI for many purposes and/or familiar with the nuanced strengths of multiple tools. You came for advanced techniques and cutting-edge applications.

Goal: Create competitive moats through advanced AI implementation

Your starting point (Month 1):

Week 1 (5-6 hours):

- Complete audit of entire tech stack (Chapter 7 comprehensive assessment)
- Identify gaps and redundancies
- Research and trial 2-3 specialized tools for your focus area
- Design end-to-end client journey automation
- Design comprehensive workflow architecture (Chapter 6 GTM engineering level)

Week 2 (5-6 hours):

- Implement complex multi-tool automation workflows
- Build data enrichment systems
- Create sophisticated lead scoring and routing logic

- Set up enterprise-level quality controls

Week 3 (5-6 hours):

- Develop proprietary systems that create competitive moats
- Build comprehensive client journey automation workflows end-to-end
- Implement advanced analytics and reporting
- Create team training systems if applicable

Week 4 (5-6 hours):

- Optimize and refine all systems
- Document everything (you're building institutional knowledge)
- Calculate comprehensive ROI across all implementations
- Plan next-level capabilities (agentic AI preparation, Chapter 9 concepts)

Success criteria: You've built sophisticated, proprietary systems that create lasting competitive advantage. You're operating at the industry frontier.

Month 2-3 focus:

- Scale successful implementations
- Experiment with cutting-edge capabilities
- Share learnings with community (you're now a resource for others)
- Consider consulting or teaching other agents

Why this works: You're not limited by basics. You're building sophisticated, differentiated systems that create lasting competitive advantage. You're operating at the frontier.

The "Good Enough" Principle

Let me share something crucial: **Perfect implementation never happens.**

You will:

- Skip steps that seem less important to you
- Modify approaches to fit your personality
- Abandon some tools that work great for others

- Discover better ways than what I recommended
- Hit obstacles I didn't anticipate
- Have weeks where you don't touch any of this

All of that is fine. Actually, it's better than fine. It's necessary.

This book isn't a script to follow precisely. It's not meant to be a boring step-by-step technical manual nor a comprehensive guide to every option. Rather, I crafted this book to be a framework to adapt to your unique situation and take the next steps to elevate your current use of AI whatever stage you are in.

Progress Over Perfection

Scenario A: Agent implements listing descriptions with AI, saves 3 hours weekly, nothing else from the book.

Result: That's 156 hours reclaimed annually. Win.

Scenario B: Agent plans comprehensive implementation, gets overwhelmed, implements nothing.

Result: Zero hours saved. Zero value delivered.

One meaningful change beats ten perfect plans.

Start small. Prove value. Expand when ready. That's the path that works.

When Things Don't Go As Planned (They Won't)

Let's talk about what happens when reality hits your perfect plans.

Common Obstacles & Grace-Filled Responses

Obstacle 1: "I tried to implement this week but got slammed with three unexpected closings."

Rigid response: "I failed. I can't do this."

Grace-filled response: "Real estate is unpredictable. I'll pick this back up next week when closings calm down. Progress isn't linear."

Action: Bookmark where you stopped. Return when capacity allows. Compressed timelines are fine.

Obstacle 2: "The AI outputs are generic. This isn't saving me time because I'm rewriting everything."

Rigid response: "AI doesn't work for me. I'm done."

Grace-filled response: "My prompts need work. Let me revisit Chapter 8's prompt engineering section and refine my approach."

Action: This is a prompt problem, not an AI limitation problem. Improve inputs. Get better outputs. Chapter 8 exists for this exact reason.

Obstacle 3: "I don't understand the technical setup for automation workflows."

Rigid response: "I'm not technical enough for this. I'll just stick to manual."

Grace-filled response: "I'm not ready for Chapter 6 level automation yet. I'll start simpler and work up to that when I have more foundation."

Action: There's no shame in not being ready for advanced techniques. Master basics first. Advanced stuff can wait.

Obstacle 4: "My metrics aren't improving like the examples in Chapter 7."

Rigid response: "This isn't working. I must be doing it wrong."

Grace-filled response: "Let me troubleshoot using Chapter 8. What's not working and why?"

Action: Metrics declining = diagnostic information, not failure. Use Chapter 8's frameworks to identify root causes and fix them.

Obstacle 5: "I implemented three things and only one is actually helping."

Rigid response: "Two out of three failed. I wasted my time."

Grace-filled response: "One out of three is delivering clear value. That's a win. I'll abandon the two that don't work and expand the one that does."

Action: Not every technique works for every agent. Discovering what DOESN'T work for you is valuable. Double down on what does.

Obstacle 6: "I can't afford the recommended tool stack right now."

Rigid response: "I can't implement this without spending money I don't have."

Grace-filled response: "I'll start with free tools and prove value before investing in premium options."

Action: Most techniques work with free ChatGPT. Premium tools optimize, but aren't mandatory for getting started. Start where your budget allows.

The pattern: Every obstacle has a rigid interpretation (that leads to quitting) and a grace-filled interpretation (that leads to adaptation).

Choose grace. Choose adaptation. Choose progress over perfection.

The Accountability System That Actually Works

Knowledge without accountability rarely becomes action. But harsh accountability creates resistance.

Here's a gentler approach:

The Weekly Check-In (5-10 Minutes)

Every Friday (or whatever day works for you), spend 5 minutes answering three questions:

1. What did I do this week related to AI?

(Even if small. Even if nothing. Just honest accounting.)

2. Did it help? How do I know?

(Time saved? Better quality? Client feedback? Measurable somehow?)

3. What's my one focus for next week?

(Not five things. ONE thing.)

Write it down. Doesn't matter where. Journal, notes app, shared doc with accountability partner, whatever. Just externalize it.

Why this works: You're tracking progress without judgment. You're staying honest without self-flagellation. You're creating gentle momentum.

The Monthly Reflection (20-30 Minutes)

Once a month, slightly deeper reflection:

1. What's working? What should I do more of?
2. What's not working? What should I stop or change?
3. What's my next capability to build?
4. Do my current efforts align with my "why" from the beginning of this chapter?

Why this works: Monthly cadence is sustainable. You're course-correcting based on real experience, not theoretical plans.

The Accountability Partner (Optional But Powerful)

Find one other agent (not in your direct market) who's also implementing AI.

Weekly 15-minute call:

- Each person shares: one win, one challenge
- You help each other troubleshoot
- You celebrate progress together
- You normalize the messy middle of implementation

Why this works: External accountability creates gentle pressure. Shared journey reduces isolation. Collaborative problem-solving accelerates both of you.

How to find one: Post in real estate AI communities (Facebook groups, LinkedIn). "Looking for accountability partner for AI implementation. 15 min weekly check-ins. Anyone interested?"

For Brokers & Team Leaders: Scaling This

If you're reading this as a broker or team leader hoping to guide multiple agents, here's adapted guidance:

The Biggest Mistakes Leaders Make

Mistake 1: Mandating Tools Without Training

"Everyone must use ChatGPT by next month."

Result: Resistance, poor implementation, abandonment

Better approach: "We're providing training on AI tools that can save you 10+ hours weekly. Participation optional but highly encouraged. Here's support available."

Mistake 2: Assuming One Size Fits All

Forcing every agent onto identical workflows and tools

Result: The approach works for some, fails for others, creates frustration

Better approach: Provide framework and support. Let agents adapt to their personalities and practices. Share what's working across the team.

Mistake 3: No Measurement or Accountability

"Try using AI sometime!"

Result: Nothing changes. Initiatives die quietly.

Better approach: "Track three metrics: time spent on listing prep, lead response time, client satisfaction. Let's review quarterly. Who's seeing improvement?"

The Tiered Adoption Approach

Tier 1: Early Adopters (10-20% of team)

- These agents are naturally tech-forward
- Give them resources, freedom to experiment, and budget support
- Task them with testing tools and sharing learnings
- They become your internal experts

Tier 2: Pragmatic Majority (60-70% of team)

- They'll adopt when value is proven
- Show them what Tier 1 agents have accomplished (metrics, testimonials)
- Provide clear, simple pathways
- Offer hands-on training and support
- Celebrate their successes publicly

Tier 3: Skeptics & Resistors (10-20% of team)

- Don't force it
- Let them see Tier 2 results over 6-12 months
- Some will eventually come around
- Some won't, and that may be okay depending on their performance

Why this works: You're working with human nature (adoption curve) not against it. You're creating positive peer pressure through demonstrated results.

The Support Structure

Monthly "AI Office Hours"

- 60-minute open session where agents can ask questions, share wins, troubleshoot problems
- Facilitated by your Tier 1 early adopters or external expert
- Recorded for agents who can't attend live

Shared Prompt Library

- Centralized doc with proven prompts for common tasks
- Agents contribute what works for them
- Regularly updated based on new discoveries

Tool Budget Guidelines

- Clear policy on what brokerage covers vs. agent-funded
- Recommended vs. required tools
- Process for requesting new tool evaluation

Quarterly Metrics Review

- Aggregate team data: time savings, efficiency gains, client satisfaction
- Share success stories
- Identify agents who need additional support

Why this works: You're providing structure and support without mandate. You're creating community learning. You're measuring what matters.

Additional online guides and training materials for brokers and team development...

https://go.howdoiai.pro/re-brokers-teams

The Final Word: Your Next Single Step

You've reached the end of this book. You have frameworks, tools, techniques, examples, troubleshooting guides, and future-proofing strategies.

But none of it matters if you don't take one action.

Not ten actions. Not a perfect comprehensive plan. **One action.**

So, here's your final assignment:

Before you close this book, commit to ONE specific action you'll take in the next 48 hours.

Write it down right now:

My one next action: _____

When I'll do it: _____

How I'll know it's complete: _____

Examples of good "one next actions":

- "Tomorrow at 9 AM, I'll spend 30 minutes creating a free ChatGPT account and generating one property description"
- "Wednesday afternoon, I'll block 1 hour to read Chapter 5 and create my first three saved prompts"
- "This weekend, I'll review my biggest time drain and choose one Chapter 6 workflow to implement next month"

Examples of bad "one next actions":

- "I'll implement AI" (too vague)
- "I'll read the book again" (more learning, not action)
- "I'll transform my entire business" (not one action, overwhelming)

Make it specific. Make it small. Make it soon.

What Success Actually Looks Like

Six months from now, if you've successfully integrated AI into your real estate practice, here's what you'll likely experience:

You won't remember most of this book. You'll have forgotten specific prompt templates, exact tool names, detailed steps. That's fine.

But you will have:

- Reclaimed 8-15 hours weekly that used to go to administrative tasks
- Built 3-5 workflows that operate reliably without constant attention
- Developed intuition for when AI helps vs. when you need human judgment
- Increased your capacity to serve clients without increasing stress
- Improved your life margin (more time for family, health, rest, relationships)

And most importantly:

- You'll have internalized the meta-skill: how to evaluate, adopt, and integrate new AI capabilities as they emerge
- You'll be navigating the years to come confidently, not fearfully
- You'll be an amplified agent, not an overwhelmed one

That's the goal. Not perfection. Not mastery of every technique. Just meaningful integration that serves your clients, your business, and your life in the ways you dream it to be.

Closing: Thank You From the Author

Dear Reader,

First and foremost, thank you for investing your valuable time in this guide. As a real estate professional, I understand that your days are filled with competing priorities—clients who need your attention, properties that need your expertise, and a business that needs your leadership.

The fact that you've chosen to explore how AI can enhance your practice speaks volumes about your commitment to growth and excellence. This willingness to embrace new tools while maintaining the human touch that defines great real estate service will set you apart in an evolving industry.

My goal in creating "How Do I AI? For Real Estate Agents" was simple: to provide a clear, practical roadmap for incorporating powerful AI capabilities into your business without sacrificing the relationship-driven foundation that makes you successful. I hope this guide has demystified AI and given you concrete steps to implement these tools in ways that amplify rather than replace your unique expertise.

Remember, this is a journey of continuous learning and refinement. The most successful agents won't be those who use every available AI tool, but those who strategically apply the right tools to enhance their distinctly human strengths.

Your Digital Resource Hub

The learning doesn't stop with the final page of this book. To support your ongoing implementation and growth, I've created a comprehensive digital resource hub exclusively for readers.

Visit https://go.howdoiai.pro/re-resources **to access:**

- Downloadable versions of all prompts in this guide

- Regularly updated tool recommendations as the landscape evolves
- Implementation templates and checklists
- Video tutorials for key AI applications
- Case studies from real estate professionals successfully using AI

Some resources are freely available to all readers, while our premium content library offers deeper implementation support for those ready to accelerate their AI journey.

Share Your Experience

Your questions, feedback, and experiences are invaluable to me and to future readers. I'm committed to making this guide as helpful as possible and to continuously improving it based on real-world implementation.

Please share your thoughts:
- Questions about specific implementations
- Success stories from your AI journey
- Challenges you've encountered
- Suggestions for future editions

https://go.howdoiai.pro/realtors-talkback

Your honest reviews also help other real estate professionals discover this resource. If you've found value in these pages, please consider sharing your experience on Amazon, Goodreads, or your preferred social platform.

For Brokerages and Teams

Implementing AI effectively across an entire organization requires strategic planning, clear guidelines, and consistent training. If you lead a brokerage or team seeking clarity and confidence in integrating AI tools throughout your organization, I'd welcome the opportunity to discuss your specific needs.

Whether you're looking to develop an AI implementation roadmap, create custom training for your team, or establish ethical guidelines for AI use in your brokerage, my consulting practice is focused on helping mid-sized service-based organizations navigate this technology transition successfully.

Request a complimentary 30-minute AI consultation specific to your business growth strategy at: https://go.howdoiai.pro/schedule-30m

The Future Is Human+AI

As we close this guide, I'd like to reiterate one central thought: The future of real estate is empowered agents using AI to deliver more value, create more impact, and build stronger relationships than ever before.

The most exciting developments aren't in what AI can do alone, but in what you can accomplish when you combine your irreplaceable human expertise with these powerful new tools.

I'm excited to see what you create on this journey.

To your continued success,

Jim Washok

AI Implementation Architect, Author, Speaker, & Workshop Facilitator

jimwashok.com

Appendix A: Prompt Library for Real Estate Professionals

This comprehensive prompt library provides ready-to-use templates for common real estate tasks. Each prompt is designed to produce high-quality results with minimal editing required. Simply copy, customize the bracketed sections with your specific information, and use with your preferred AI assistant.

How to Use This Prompt Library

This appendix contains ready-to-use prompt templates for the most common real estate tasks covered throughout this book. Each prompt is designed to produce professional-quality results with minimal editing—but only if you customize it properly.

Three critical points before you begin:

1. These are starting points, not finished products. Copy any prompt, replace the [BRACKETED] sections with your specific information, and submit to your preferred AI assistant (ChatGPT, Claude, etc.). Review the output. Refine your prompt based on what you get. Save your best-performing versions. The process is iterative, not one-and-done.

2. Add your brand voice for consistency. For best results, paste your brand voice guide (see Chapter 8's template) before or alongside these prompts. This ensures AI outputs match your personal communication style rather than sounding generic. Without your voice context, even great prompts produce bland results.

3. These prompts work with any major LLM. You're not locked into a specific tool. Use these with ChatGPT, Claude, Gemini, or whatever AI platform you prefer. The principles of effective prompting (covered in Chapter 2) apply universally.

Don't try to use all of these at once. Pick 2-3 prompts that address your biggest current needs. Master those. Expand from there. Quality implementation of a few beats mediocre execution of many.

Now, dive in and start reclaiming your time.

Access or download the complete library of all prompts in this book...

http://go.howdoiai.pro/re-all-prompts-e1

Property Description Prompts

Standard Residential Listing Description

> Create a compelling property description for a [PROPERTY TYPE] home in [NEIGHBORHOOD/CITY].

Key details:

- [X] bedrooms, [X] bathrooms
- [X] square feet
- Built in [YEAR]
- [LIST 3-5 STANDOUT FEATURES]
- [NOTABLE ARCHITECTURAL STYLE OR UPDATES]
- Located near [NEIGHBORHOOD AMENITIES]

Target buyers are likely [DESCRIBE IDEAL BUYER PROFILE].

The tone should be
[ELEGANT/ENTHUSIASTIC/PROFESSIONAL] and
approximately 250-300 words.

Highlight these key selling points:

- [PRIORITY FEATURE 1]

- [PRIORITY FEATURE 2]

- [PRIORITY FEATURE 3]

Avoid these terms: [ANY TERMS TO AVOID FOR FAIR
HOUSING OR PREFERENCE]

Luxury Property Description

> Create a sophisticated, emotionally compelling description for a
luxury [PROPERTY TYPE] in [LOCATION].

Property highlights:

- [X] bedrooms, [X] bathrooms, [X] square feet

- Built/Renovated: [YEAR]

- Architect/Designer: [IF APPLICABLE]

- Lot size: [SIZE]

Signature features:

- [LUXURY FEATURE 1]

- [LUXURY FEATURE 2]

- [LUXURY FEATURE 3]

- [LUXURY FEATURE 4]

- [LUXURY FEATURE 5]

Location advantages:

- [LOCATION BENEFIT 1]

- [LOCATION BENEFIT 2]

- [LOCATION BENEFIT 3]

The ideal tone is sophisticated and aspirational while remaining authentic. Focus on craftsmanship, uniqueness, and the lifestyle the property enables rather than just amenities.

Length: Approximately 350-400 words.

Include a powerful opening paragraph that captures the essence of the property and a strong closing statement emphasizing exclusivity without using potentially discriminatory language.

Investment Property Description

> Create a fact-based, ROI-focused property description for a [PROPERTY TYPE] investment opportunity in [LOCATION].

Property specifications:

- [X] units/bedrooms, [X] bathrooms

- [X] square feet total

- Built in [YEAR], [ANY MAJOR RENOVATIONS]

- Current rental income: $[AMOUNT]/month

- Current expenses: $[AMOUNT]/month

- Cap rate: [PERCENTAGE]

Key investment advantages:

- [ADVANTAGE 1]

- [ADVANTAGE 2]

- [ADVANTAGE 3]

- [ADVANTAGE 4]

Market highlights:

- [MARKET TREND 1]

- [MARKET TREND 2]

- [MARKET TREND 3]

The tone should be professional and data-driven, appealing to investment-minded buyers. Focus on financial performance, upside potential, and market stability factors.

Length: Approximately 250-300 words.

Avoid making specific return guarantees while still highlighting the property's investment merits.

Fixer-Upper/Renovation Opportunity

> Create an honest yet opportunity-focused property description for a [PROPERTY TYPE] fixer-upper in [LOCATION].

Property basics:

- [X] bedrooms, [X] bathrooms

- [X] square feet

- Built in [YEAR]

- Current condition: [BRIEF ASSESSMENT]

Renovation potential:

- [OPPORTUNITY 1]

- [OPPORTUNITY 2]

- [OPPORTUNITY 3]

- [OPPORTUNITY 4]

Location benefits:

- [BENEFIT 1]

- [BENEFIT 2]

- [BENEFIT 3]

The tone should balance honesty about the current condition with enthusiasm about the potential. Target audience is [INVESTORS/DIY HOMEBUYERS/CONTRACTORS].

Length: Approximately 250-300 words.

Highlight the value proposition clearly while being forthright about the work required. Focus on the end result that could be achieved with appropriate investment.

Client Communications Prompts

Initial Buyer Consultation Follow-Up

> Write a warm, professional follow-up email to send after an initial buyer consultation meeting.

Client details:

- Names: [CLIENT NAMES]

- Looking for: [PROPERTY TYPE] in [LOCATION(S)]

- Budget range: [BUDGET]

- Timeline: [TIMELINE]

- Key requirements: [2-3 MUST-HAVES]

- Specific concerns raised: [ANY CONCERNS MENTIONED]

Include:

1. A personalized thank you for meeting

2. A brief summary of what we discussed about their needs

3. Confirmation of next steps we agreed to

4. 2-3 initial properties that might interest them based on our conversation

5. A clear call to action for scheduling our next communication

6. My availability for the coming week

The tone should be professional yet conversational and approachable. Length should be approximately 250-300 words.

Include [BRACKETS] for information I need to customize for each client.

Listing Update for Sellers

> Create a market update email for a seller client whose property has been on the market for [TIME PERIOD].

Property details:

- Address: [ADDRESS]

- List price: [PRICE]

- Days on market: [DAYS]

- Showing activity: [NUMBER OF SHOWINGS]

- Feedback themes: [COMMON FEEDBACK]

- Market changes since listing: [MARKET CHANGES]

Include:

1. A supportive opening acknowledging their partnership

2. Factual data about market activity and their specific listing

3. Summary of feedback received from showings

4. Analysis of how their property compares to recent sales/competition

5. Recommended adjustments if any (pricing, presentation, etc.)

6. Clear next steps and timeline

7. Reassurance and availability for questions

The tone should be straightforward but supportive, avoiding minimizing concerns while maintaining confidence.

Length: Approximately 300-350 words.

Include [BRACKETS] for information I need to customize.

Offer Presentation Email

> Write an email to my seller client presenting an offer we've received on their property.

Offer details:

- Property address: [ADDRESS]

- List price: [LIST PRICE]

- Offer price: [OFFER PRICE]

- Down payment: [AMOUNT]

- Loan type: [LOAN TYPE]

- Closing timeline: [TIMELINE]

- Key contingencies: [LIST CONTINGENCIES]

- Notable terms: [ANY UNUSUAL TERMS]

- Buyer profile: [BRIEF BUYER DESCRIPTION WITHOUT FAIR HOUSING ISSUES]

Include:

1. A clear subject line indicating an offer has been received

2. Brief introduction of the offer's key points

3. Summary of financial terms

4. Summary of key contingencies and timeline

5. Initial analysis of offer strengths and considerations

6. Proposed time for us to discuss in detail

7. Reminder of our process for reviewing offers

8. My immediate availability for questions

The tone should be professional and objective, presenting information without pushing the client toward a specific decision.

Length: Approximately 250-300 words.

Include [BRACKETS] for information I need to customize.

Past Client Check-In

> Create a personalized check-in email for a past client who purchased their home [TIME PERIOD] ago.

Client details:

- Names: [CLIENT NAMES]

- Property: [PROPERTY TYPE] in [NEIGHBORHOOD]

- Purchase date: [DATE]

- Any life updates I know: [UPDATES IF ANY]

- Local market changes: [BRIEF MARKET UPDATE]

- Their interests: [ANY KNOWN INTERESTS]

Include:

1. A warm, genuine opening referencing our past work together

2. Brief, valuable insight about their neighborhood market

3. Non-intrusive question about how they're enjoying the home

4. Useful local information (event, restaurant opening, community news)

5. Subtle reminder of my continued service and referral business

6. No-pressure invitation to connect

7. Personal closing note

The tone should be friendly and service-oriented without being sales-focused. This should feel like a check-in from a caring professional, not a solicitation.

Length: Approximately 200-250 words.

Include [BRACKETS] for information I need to customize.

Market Analysis Prompts

Neighborhood Market Update

> Create a data-driven neighborhood market update for [NEIGHBORHOOD] in [CITY], focusing on [PROPERTY TYPE] homes.

Include these data points with analysis:

- Current active listings: [NUMBER]

- Average days on market: [NUMBER] (compared to [PREVIOUS PERIOD])

- Median sale price: [PRICE] (compared to [PREVIOUS PERIOD])

- Price per square foot: [AMOUNT]

- Months of inventory: [NUMBER]

- Absorption rate: [RATE]

- List-to-sale price ratio: [PERCENTAGE]

Recent market developments:

- [DEVELOPMENT 1]

- [DEVELOPMENT 2]

- [DEVELOPMENT 3]

The tone should be factual and analytical while remaining accessible to non-real estate professionals. Explain what these metrics mean for [BUYERS/SELLERS] in plain language.

Format with clear headings, concise paragraphs, and a brief "Market Implications" summary at the end.

Length: Approximately 400-450 words.

Comparative Market Analysis Framework

> Create a structured framework for presenting a Comparative Market Analysis to a [SELLER/BUYER] client for a [PROPERTY TYPE] in [NEIGHBORHOOD].

Subject property details:

- [X] bedrooms, [X] bathrooms

- [X] square feet

- Built in [YEAR]

- Key features: [LIST FEATURES]

- Condition: [CONDITION]

Data to include:

1. Active competition analysis

2. Recent sold comparables analysis

3. Expired/withdrawn listings insights

4. Market trend indicators for this specific segment

5. Absorption rate and days on market analysis

6. Pricing strategy recommendations

For each section, provide:

- What specific data to present

- How to visually present it (chart type, format)

- Key talking points to explain significance

- Questions to engage the client

The framework should be designed as a conversation guide, not just a data dump. Include specific transition statements between sections and a clear pricing recommendation approach.

Length: Approximately 500-600 words outlining the complete presentation framework.

Investment Property Analysis

> Create an investment property analysis for a [PROPERTY TYPE] in [LOCATION] being considered for [LONG-TERM RENTAL/SHORT-TERM RENTAL/FIX-AND-FLIP].

Property details:

- Purchase price: $[AMOUNT]

- Estimated repairs/updates: $[AMOUNT]

- Square footage: [SQFT]

- Bedrooms/Bathrooms: [BED]/[BATH]

- Year built: [YEAR]

- Current condition: [CONDITION]

For a rental analysis, include:

- Estimated monthly rent: $[AMOUNT]

- Estimated vacancy rate: [PERCENTAGE]

- Property management cost: [PERCENTAGE or AMOUNT]

- Property tax: $[AMOUNT]/year

- Insurance estimate: $[AMOUNT]/year

- Maintenance reserve: [PERCENTAGE]

- HOA fees (if applicable): $[AMOUNT]

- Utilities paid by owner (if any): $[AMOUNT]

Calculate and explain:

1. Cash flow analysis (monthly and annual)

2. Cash-on-cash return

3. Cap rate

4. Gross rent multiplier

5. Estimated appreciation based on local trends

6. Total ROI projection (1-year and 5-year)

7. Break-even analysis

For fix-and-flip, also include:

- Expected timeline

- After-repair value (ARV)

- Profit potential

- Return on investment

Present this data in a clear, analytical format with explanations of what each metric means for the investment decision. Include a summary recommendation section highlighting the key factors to consider.

Length: Approximately 500-550 words.

School District Analysis

> Create a comprehensive, fair-housing compliant school district analysis for families considering [NEIGHBORHOOD / AREA].

Include information on:

Public schools serving the area:

- Elementary: [SCHOOL NAMES]

- Middle: [SCHOOL NAMES]

- High: [SCHOOL NAMES]

For each school, provide:

- School size and student-teacher ratios

- Notable programs (STEM, arts, languages, etc.)

- Extracurricular offerings

- Facilities information

Also include:

- Private school options within [X] miles

- Preschool and daycare options

- Enrichment programs and educational resources

- Community education programs

- School district funding and budget information

- Recent developments or changes in the district

Present the information in an objective, data-focused manner without making subjective quality judgments. Include links/references to where families can research school performance data on their own.

Compliance note: Ensure the analysis focuses on factual information about educational options without steering or making value judgments that could violate fair housing laws.

Length: Approximately 450-500 words.

Transaction Management Prompts

Transaction Timeline Creation

> Create a detailed transaction timeline for a [BUYER/SELLER] in [STATE] with a closing date of [DATE].

Transaction details:

- Contract acceptance date: [DATE]

- Property type: [TYPE]

- Financing type: [CASH/CONVENTIONAL/FHA/VA]

- Contingencies: [LIST CONTINGENCIES]

- Special circumstances: [ANY SPECIAL CIRCUMSTANCES]

The timeline should:

1. List all key deadlines chronologically with specific dates

2. Indicate responsible parties for each task

3. Differentiate between contractual deadlines and recommended actions

4. Include preparation steps before major milestones

5. Note document submission requirements and review periods

6. Include post-closing actions and recommendations

Format as a day-by-day calendar with clear categorization of:

- Critical contractual deadlines (highlighted)

- Recommended preparation steps

- Communication checkpoints

- Document collection and submission

- Coordination with third parties

Include explanations of why each step matters and potential consequences of missed deadlines.

Length: Approximately 500-600 words covering the entire transaction period.

Inspection Response Strategy

> Develop a strategic framework for responding to a home inspection report for my [BUYER/SELLER] client.

Inspection findings overview:

- Major issues: [LIST MAJOR ISSUES]

- Safety concerns: [LIST SAFETY ITEMS]

- Maintenance items: [LIST MAINTENANCE ITEMS]

- Approximate cost to address all items: $[ESTIMATED TOTAL]

- Transaction context: [MARKET CONDITIONS, MULTIPLE OFFERS, ETC.]

For buyer client strategy, include:

1. Prioritization framework for requests (safety, functionality, cost)

2. Options for different request approaches (repair, credit, price reduction)

3. Strategic considerations based on market conditions

4. Communication approach for presenting requests

5. Fallback positions and negotiation flexibility

For seller client strategy, include:

1. Analysis of likely buyer concerns and priorities

2. Proactive response options (immediate repairs, inspection credits)

3. Negotiation parameters and flexibility considerations

4. Cost-benefit analysis of different response approaches

5. Risk assessment of potential outcomes

The guidance should be strategic rather than purely technical, focusing on negotiation positioning and transaction management rather than the repairs themselves.

Length: Approximately 400-450 words of actionable strategic guidance.

Closing Preparation Checklist

> Create a comprehensive closing preparation checklist for a [BUYER/SELLER] with a closing date of [DATE] in [STATE].

Transaction details:

- Property address: [ADDRESS]

- Closing location: [LOCATION]

- Closing method: [IN-PERSON / REMOTE]

- Special circumstances: [ANY SPECIAL CIRCUMSTANCES]

For a buyer checklist, include:

1. Final financing requirements and deadlines

2. Insurance procurement steps and documentation

3. Utility transfer procedures and contacts

4. Final walkthrough preparation and checklist

5. Closing funds preparation with timeline and method

6. Required identification and documentation

7. Post-closing immediate actions

For a seller checklist, include:

1. Property preparation requirements

2. Required repair completion and documentation

3. Property vacancy timeline and requirements

4. Document collection and submission requirements

5. Utility management and transfer procedures

6. Closing proceeds instructions and options

7. Moving coordination and property handover steps

Format as a chronological checklist with timing indicators (e.g., "7 days before closing," "3 days before closing," "Closing day") and clear responsible parties for each action.

Include explanatory notes for first-time buyers / sellers and specific local requirements for this jurisdiction.

Length: Approximately 450-500 words in a structured checklist format.

Multiple Offer Analysis

> Create a structured analysis framework for presenting multiple offers on my seller's property.

Offer overview:

- Number of offers received: [NUMBER]

- List price: $[AMOUNT]

- Property address: [ADDRESS]

- Days on market: [DAYS]

- Special considerations: [ANY SPECIAL CIRCUMSTANCES]

For each offer, create a comparison framework including:

1. Offer price and net proceeds calculation

2. Financing type and down payment amount

3. Pre-approval/proof of funds strength assessment

4. Contingencies included and their timelines

5. Proposed closing timeline

6. Unusual terms or requests

7. Buyer motivation information (if known)

8. Likelihood of closing assessment

Also include:

- Side-by-side comparison matrix of all key terms

- Risk assessment for each offer

- Potential counter-offer strategies

- Multiple offer response options (accept, counter, multiple counter)

The analysis should be objective and thorough, presenting options rather than pushing for a specific decision. Include strategic considerations beyond just price.

Length: Approximately 500-600 words plus comparison matrix.

Business Development Prompts

Listing Presentation Outline

> Create a comprehensive outline for a listing presentation for a [PROPERTY TYPE] in [NEIGHBORHOOD] valued at approximately $[AMOUNT].

Market context:

- Current inventory levels: [LEVEL]

- Average days on market: [DAYS]

- Recent pricing trends: [TREND]

- Notable market factors: [FACTORS]

The presentation outline should include these sections with specific talking points:

1. Introduction and rapport building

   - Personal connection points

   - Listening framework for their specific needs

   - Transition to presentation

2. Market overview

   - Current conditions specific to their property type

   - Supply and demand analysis

   - Buyer profile and expectations

   - Seasonal considerations

3. Property valuation

   - Comparative market analysis approach

   - Value-adding features assessment

   - Pricing strategy options

   - Absorption rate analysis

4. Marketing strategy

   - Property preparation recommendations

   - Photography and visual marketing

- Online and offline marketing channels

- Target marketing to qualified buyers

- Showing strategy

5. Transaction management

- Communication expectations

- Showing feedback system

- Offer evaluation process

- Negotiation approach

- Closing coordination

6. Unique value proposition

- Specific differentiators from competitors

- Past success stories relevant to their situation

- References and testimonials

- Team/resources available

7. Next steps and timeline

- Immediate action items

- Preparation schedule

- Marketing launch timing

- Expectation setting

Include engagement questions for each section and visual elements to incorporate. Format as a structured outline with bullet points and presenter notes.

Length: Approximately 600-700 words covering the complete presentation framework.

Real Estate Newsletter Content

> Create content for a monthly real estate newsletter for my sphere of influence in [LOCATION].

Newsletter sections should include:

1. Market update (approximately 150 words)
   - Current inventory levels
   - Median sales price trends
   - Days on market changes
   - Interest rate impact
   - Seasonal factors

2. Featured neighborhood spotlight on [NEIGHBORHOOD] (approximately 200 words)
   - Brief history or interesting facts
   - Current market snapshot
   - Notable amenities or developments
   - Community events or highlights
   - Investment perspective

3. Homeowner tip of the month: [SEASONAL TOPIC] (approximately 150 words)
   - Practical, actionable advice
   - Estimated cost/time investment
   - Professional resources if needed
   - Long-term benefit explanation

4. Real estate Q&A (approximately 100 words)
   - Question: [COMMON CLIENT QUESTION]
   - Clear, concise expert answer
   - Additional resource recommendation

5. Personal/business update (brief section for personalization)
   - Recent business milestone or client story (no names)
   - Community involvement

- Brief personal note

The tone should be informative yet conversational, positioning me as a knowledgeable resource rather than being sales-focused. Include a subtle call to action for referrals or consultation requests.

Length: Approximately 650-700 words total for the complete newsletter.

Social Media Content Calendar

> Create a 2-week social media content calendar for a real estate agent in [LOCATION] focusing on [NICHE/SPECIALTY].

For each post, include:

1. Platform (Instagram, Facebook, LinkedIn)

2. Content type (image, video, carousel, story)

3. Complete post copy including hook, body, and call to action

4. Hashtag recommendations (5-7 relevant tags)

5. Optimal posting time

6. Visual content suggestions

Content mix should include:

- 2 market update posts with actionable insights

- 2 listing/property highlight posts

- 2 educational posts about the buying/selling process

- 1 neighborhood spotlight

- 1 client success story framework (no specific client details)

- 1 personal branding/behind-the-scenes content

- 1 seasonal or timely topic

The tone should be [PROFESSIONAL/CONVERSATIONAL/ENERGETIC] and aligned with my brand voice. Each post should provide actual value to followers rather than just promotional content.

Include engagement prompts and questions to encourage interaction. Ensure all content is fair housing compliant and ethically sound.

Length: Calendar covering 10 posts with approximately 100-150 words per post description.

Referral Request Sequence

> Create a 3-part email sequence to request referrals from past clients and sphere of influence.

Client relationship context:

- Transaction type: [BUYER/SELLER/BOTH]

- Time since transaction: [TIME PERIOD]

- Communication frequency: [REGULAR/OCCASIONAL/LIMITED]

- Special circumstances: [ANY NOTABLE FACTORS]

Email 1: Value-First Reconnection

- Subject line options (provide 3)

- Opening that references our past work without being transactional

- Valuable market insight relevant to their property/area

- Personal check-in element

- Subtle mention of business growth through referrals

- No direct ask in this email

- Length: Approximately 200-250 words

Email 2: Educational Content + Soft Ask (to be sent 5-7 days later)

- Subject line options (provide 3)

- Reference to previous email

- Useful education content about [CURRENT MARKET TOPIC]

- Explanation of who makes an ideal referral for my business

- Soft, no-pressure referral request

- Simple response mechanism

- Length: Approximately 250-300 words

Email 3: Direct Request + Incentive (to be sent 7-10 days after Email 2)

- Subject line options (provide 3)

- Clear, confident referral request

- Specific explanation of how I help referred clients

- Process explanation (what happens when they refer someone)

- Appropriate incentive or thank you gesture

- Clear call to action with multiple response options

- Length: Approximately 200-250 words

The tone should be authentic and relationship-focused throughout, not sales-driven or desperate. Include personalization placeholders in [BRACKETS].

Length: Three complete emails with subject line options for each.

This prompt library provides a strong foundation for leveraging AI across your real estate business. For best results, customize each prompt with your specific information and brand voice before submitting to your AI assistant. Save successful prompts and continue refining them based on the results you receive.

Appendix B: Tool Comparison Matrix

This matrix provides a side-by-side comparison of key AI tools for real estate professionals, helping you select the right solutions for your specific needs and budget. Tools are organized by category with primary features, pricing, and considerations highlighted.

How to use this appendix: Start by identifying your biggest pain point (content creation? lead follow-up? property marketing?), then review the relevant category. Compare tools based on your budget and technical comfort level. Cross-reference the chapter where each tool is discussed in detail for implementation guidance.

Prices, features, limitations, URLs, and available tools subject to change by vendors without notice. Be sure to verify information on vendor websites before making purchase decisions.

General-Purpose AI Assistants

Tool	Best For	Key Features	Limitations	Pricing (2025)	Chapter
Chat-GPT Plus	Quick content creation, market analysis, versatile daily use	• Fast response times • Image understanding • Web browsing for research • Canvas for collaborative editing • Voice interaction	• Can be verbose • Sometimes overly formal • May miss nuance in complex instructions	$20/ month	Ch 2, 4, 5

Tool	Best For	Key Features	Limitations	Pricing (2025)	Chapter
Claude (Sonnet /Opus)	Long-form content, complex analysis, nuanced tone matching	• 200K token context window • Superior instruction following • More natural writing style • Better at matching specific voices • Excellent for complex reasoning	• Slower response times • Fewer integrations • More expensive for API use	$20/month (Pro) Free tier available	Ch 4, 5, 8
Perplexity Pro	Market research, current data, citation-heavy analysis	• Real-time web search • Source citations • Academic research mode • Multiple AI models • Current market data access	• Less creative writing • Not ideal for tone matching • Better for research than content	$20/month Free tier available	Ch 5, 7
Gemini Adv.	Google ecosystem users, multi-modal tasks	• Deep Google integration • Strong at data analysis • Multimodal capabilities • Google Workspace integration	• Less sophisti-cated writing • UI learning curve • Fewer real estate use cases documented	$20/month Free tier available	Ch 4

Recommendation: Start with ChatGPT Plus for versatility. Add Claude Pro if you need sophisticated long-form content (luxury listings, detailed market reports, complex client communications). Add Perplexity Pro if market research is a major time drain.

Property Marketing Tools

Tool	Best For	Key Features	Limitations	Pricing (2025)	Chapter
Virtual Staging AI	Professional staging for empty properties	• Multiple design styles • Realistic furniture placement • Quick turnaround • High quality results • Renovation visualization	• Works best with empty rooms • Occasional artifacts • Per-room pricing on basic plan	$12/month basic $29/month premium $79/month professional	Ch 5
Reimagine-home	Basic photo enhancements, sky replacement, quick edits	• Free tier available • Sky replacement • Item removal • Lawn enhancement • Simple interface	• Quality varies • Limited staging sophistication • Free tier restricts monthly use	Free for 30 images/month $19/month premium	Ch 5
Canva Pro	Marketing materials, social media, presentations	• AI-powered design (Magic Design) • Real estate templates • Brand kit storage • Background remover • Video	• Learning curve for advanced features • Some AI features require premium • Not real estate specific	$119/year ($10/month equivalent)	Ch 5

Tool	Best For	Key Features	Limitations	Pricing (2025)	Cha pter
		editing capabilities			
Adobe Express	Profess-ional-grade design, brand consistency	• AI design generation • Advanced editing tools • Adobe Creative Cloud integration • Professional templates	• Higher cost • Steeper learning curve • May be overkill for many agents	$99/ year Included with Creative Cloud	Ch 5

Recommendation: Every agent should have either Virtual Staging AI (if you list properties) or Reimaginehome free tier (occasional use). Add Canva Pro for comprehensive marketing materials.

Workflow Automation & Integration

Tool	Best For	Key Features	Limitations	Pricing (2025)	Cha pter
Zapier	Simple auto-mation, connecting existing tools	• 7,000+ app inte-grations • No-code automation • Pre-built templates • Reliable execution • Good docu-mentation	• Task-based pricing adds up • Limited data transfor-mation • Not designed for complex GTM workflows	Free tier: 100 tasks/ month $19.99/ month: 750 tasks $49/ month: 2,000 tasks	Ch 6
Make	Complex work-flows,	• Visual workflow builder	• Steeper learning curve	Free tier: 1,000 operations/	Ch 6

Tool	Best For	Key Features	Limitations	Pricing (2025)	Chapter
	GTM engineering, data transformation	• Advanced data manipulation • More affordable than Zapier • Better for complex logic • Powerful filtering and routing	• Fewer pre-built templates • Less user-friendly for beginners	month $9/month: 10,000 operations $16/month: 10,000+ operations	
Clay	GTM engineering, data enrichment, investor outreach	• Multi-source data enrichment • AI-powered personalization • Waterfall enrichment • Spreadsheet-like interface • LinkedIn integration	• Credit-based pricing • Learning curve • Best for high-volume outreach • Requires strategic setup	Free tier: 100 credits $149/month: 3,000 credits $349/month: 10,000 credits	Ch 6

Recommendation: Start with Zapier if you're new to automation. Graduate to Make if you need complex workflows or have budget constraints. Add Clay only if you're doing sophisticated GTM engineering (investor outreach, FSBOs at scale).

Voice & Field AI Tools

Tool	Best For	Key Features	Limitations	Pricing (2025)	Chapter
Fixer	Voice-to-CRM updates, field data capture	• Voice note transcription • Direct CRM integration • Task creation from voice • Mobile-first design • Real-time processing	• CRM integration varies • Requires clear speech • May need WiFi/data • Limited to supported CRMs	Custom pricing (typically $50-100/month)	Ch 3, 6
House-Whisper	Showing notes, client meeting summaries	• Real estate specific • Meeting transcription • Client profile updates • Automated follow-up suggestions	• Real estate focused only • Newer platform • Integration ecosystem growing	Starting at $49/month	Ch 3
Native Voice Tools (Chat-GPT Voice, Claude voice features)	Quick content creation on the go	• No additional cost • Built into LLM apps • Natural conversation • Multi-language support	• No CRM integration • Manual transfer required • Requires copy/paste workflow	Included with AI subscriptions	Ch 3, 5

Recommendation: If you spend significant time in the field and struggle with CRM data entry, invest in Fixer or HouseWhisper. Otherwise, start with native voice features in ChatGPT/Claude for content creation on the go.

Lead Generation & Nurturing

Tool	Best For	Key Features	Limitations	Pricing (2025)	Chapter
Roof AI	Website chatbot, 24/7 lead engagement	• Real estate trained AI • Lead qualification • Appointment setting • Customizable conversation flows • Multi-language support	• Setup time investment • Requires website integration • Ongoing optimization needed	Custom pricing (typically $200-400/month)	Ch 6
Structurely	AI-powered lead follow-up via text/email	• Automated lead response • Qualification conversations • CRM integration • Multi-touch sequences	• Can feel automated • Requires careful setup • Best for high lead volume	Custom pricing (typically $300-500/month)	Ch 6

Recommendation: Only invest in specialized lead tools if you generate 30+ leads monthly. Otherwise, use ChatGPT/Claude with automation tools (Zapier/Make) for follow-up sequences.

AI-Enhanced CRM Systems

Tool	Best For	Key Features	Limitations	Pricing (2025)	Cha pter
Follow Up Boss	Stream-lined lead manage-ment, simple auto-mation	• Intelligent lead routing • Smart action plans • Text/email integration • Accountability tools • Clean interface	• Less AI sophisti-cation • Limited predictive functions • Focused on follow-up only	$69-499/ month (based on users)	Ch 6
Lofty AI	Full AI-powered CRM for teams	• Behavioral monitoring • Adaptive lead nurturing • Customizable AI workflows • Predictive analytics • Full transaction management	• Significant investment • Complex implement-ation • Best for teams • Requires commitment	Starting at $499/ month (varies by team size)	Ch 6
Real Geeks + Geek AI	Website + CRM combo with AI chatbot	• IDX website included • AI chatbot • Lead generation tools • Basic automation • SEO features	• Ecosystem lock-in • Less customiz-able AI • Best for full adoption	Starting at $299/ month (varies by features)	Ch 6

Recommendation: Most solo agents don't need to switch CRMs for AI features. Use your existing CRM + external AI tools (ChatGPT + Zapier/Make) for 90% of the benefit at 10% of the cost.

Compliance & Security Tools

Tool	Best For	Key Features	Limitations	Pricing (2025)	Chapter
Snappt	Fraud detection for rental applications	• 99.8% fraud detection accuracy • Automated document verification • Paystub and bank statement analysis • Fast processing • Risk scoring	• Rental-focused only • Per-screening cost • Best for property managers • Limited sales transaction use	Custom pricing (typically $3-8 per screening)	Ch 6, 8
AI Prompt Guardrails (DIY approach)	Fair Housing compliance in AI outputs	• No additional cost • Customizable to your needs • Works with any AI tool • Full control	• Requires manual implementation • Must maintain vigilance • No automated enforcement	Free (DIY)	Ch 1, 6, 8

Recommendation: Snappt is valuable for agents managing rentals. For sales transactions, implement Fair Housing guardrails in your prompts (see Chapter 6) rather than paying for specialized tools.

Market Analysis & Data Tools

Tool	Best For	Key Features	Limitations	Pricing (2025)	Chapter
House-Canary	Property valuation, investment analysis	• AI-powered valuations • Future value projections • Risk assessment • Rental analysis • Market forecasting	• Enterprise pricing • Data quality varies by market • Overkill for most agents	Custom pricing (typically $200-500/month)	Ch 5, 7
Redfin Data Center	Free market trend research	• Free comprehensive data • Interactive visualizations • National and local insights • Trend analysis • Downloadable reports	• Limited customization • Data coverage varies • No direct integration • No predictive features	Free	Ch 5
Your MLS + AI Analysis (DIY approach)	Custom market analysis with your data	• Uses your actual market data • Fully customizable • No additional cost • Most relevant to your area	• Requires manual data export • You do the analysis • No automation • Time investment required	Free (DIY with ChatGPT/Claude)	Ch 5, 7

Recommendation: Start with free tools (Redfin Data Center + MLS + ChatGPT analysis). Only invest in HouseCanary if you work heavily with investors and need sophisticated valuation models.

For the latest updates to the tool matrix, visit

http://go.howdoiai.pro/re-tool-matrix

Comparison By Use Case

Solo Agent on Limited Budget ($20-50/month)

Recommended Stack:

- **ChatGPT Plus** ($20/month) - Core AI assistant for all content
- **Reimaginehome** (Free tier) - Basic property enhancements
- **Canva** (Free tier) - Basic marketing materials
- **Native voice features** (Included) - Content creation on the go
- **Existing CRM** with manual implementation of AI outputs

Total Monthly Investment: $20

What you get: Core content creation efficiency, basic property marketing, DIY automation through manual AI workflows.

Growing Individual Agent ($100-200/month)

Recommended Stack:

- **ChatGPT Plus** ($20/month) - Daily content creation
- **Claude Pro** ($20/month) - Sophisticated long-form content

- **Virtual Staging AI** ($12/month) - Property visualization
- **Canva Pro** ($10/month) - Professional marketing
- **Zapier** ($20/month) - Basic automation
- **Reimaginehome Premium** ($19/month) - Photo enhancements

Total Monthly Investment: $101

What you get: Comprehensive content creation, professional property marketing, basic workflow automation, significant time savings.

Established Agent with Volume ($300-500/month)

Recommended Stack:

- **ChatGPT Plus** ($20/month) - Quick content
- **Claude Pro** ($20/month) - Premium content
- **Perplexity Pro** ($20/month) - Market research
- **Virtual Staging AI Premium** ($29/month) - High-volume staging
- **Canva Pro** ($10/month) - Marketing materials
- **Make** ($16/month) - Complex automation
- **Clay** ($149/month) - GTM engineering for investor outreach
- **Fixer or HouseWhisper** ($50-100/month) - Field AI

Total Monthly Investment: $314-364

What you get: Complete AI-powered infrastructure, sophisticated automation, field integration, competitive moat through GTM engineering.

Team or High-Volume Operation ($800-1,500/month)

Recommended Stack:

- **All tools from Established Agent stack** ($314-364/month)
- **Follow Up Boss or Lofty AI** ($300-499/month) - Team CRM with AI
- **Roof AI** ($300/month) - 24/7 lead engagement
- **HouseCanary** ($300/month) - Advanced valuation for investor clients

Total Monthly Investment: $1,214-1,463

What you get: Enterprise-level AI infrastructure, team coordination tools, advanced lead generation, sophisticated data analysis, full automation.

Implementation Considerations

Factor	Questions to Consider	Tips
Integration	• Does it connect with your existing tools? • How much manual transfer is required? • Are there API capabilities?	• Prioritize tools that connect to your primary CRM • Consider Make or Zapier for custom connections • Test integration during trial period • Budget time for integration setup
Training Time	• How long until proficiency? • Is training provided? • Are there templates/presets?	• Calculate true cost including learning time • Start with user-friendly tools • Budget 2-3x listed setup time • Use Chapter 10's implementation timelines
Support	• What support is available? • Are there user communities? • How responsive is the company?	• Check support hours and methods • Join user groups for peer support • Test support during trial • Read recent reviews about support quality
ROI Timeline	• When will you recoup costs? • What's the productivity increase? • How does it impact client experience?	• Set specific KPIs for each tool (Chapter 7) • Track time savings and quality improvements • Review ROI quarterly • Kill tools that don't deliver value

Factor	Questions to Consider	Tips
Data Security	• How is client data protected? • Is it GDPR/privacy compliant? • Can you control data usage? • What's the company's track record?	• Never upload confidential client data to free AI tools • Review privacy policies • Use business/enterprise tiers for sensitive data • Follow Chapter 8 security protocols

Strategic Upgrade Pathways

For successful implementation, follow these progressive pathways based on your priority:

Foundation Path (Every Agent)

Month 1: ChatGPT Plus

Month 2: Basic photo enhancement (Reimaginehome free)

Month 3: Canva Pro for marketing

Month 4: Add Claude Pro for sophisticated content

Month 6: Basic automation (Zapier)

Goal: Core efficiency gains across content creation and property marketing.

Lead Generation Path

Month 1: ChatGPT Plus

Month 2: Zapier for follow-up automation

Month 3: Evaluate if lead volume justifies Roof AI

Month 6: Add Clay if doing investor outreach at scale

Goal: Systematic lead nurturing without dropping opportunities.

Marketing Excellence Path

Month 1: ChatGPT Plus

Month 2: Virtual Staging AI

Month 3: Canva Pro

Month 4: Claude Pro for luxury/complex listings

Month 6: Advanced design tools if needed

Goal: Best-in-market property presentation and brand presence.

GTM Engineering Path (Advanced)

Month 1-3: Foundation Path tools

Month 4: Make for workflow automation

Month 5: Clay for data enrichment

Month 6: Build sophisticated investor/FSBO outreach systems

Goal: Competitive moat through sophisticated automation unavailable to typical agents.

Key Decision Framework

When evaluating any new AI tool, use this framework:

Question 1: What specific problem does this solve?

If you can't articulate a clear problem, don't buy the tool.

Question 2: What's my current workaround costing me?

Calculate time or money spent on the problem. Tool must cost less.

Question 3: Can I solve this with tools I already have?

Often ChatGPT + Zapier solves problems that specialized tools claim to fix.

Question 4: Will I use this consistently?

Be honest. Shiny objects that sit unused waste money.

Question 5: Can I test it with low commitment?

Free trials, monthly plans, and money-back guarantees reduce risk.

If you answer these five questions positively, pilot the tool using Chapter 4's evaluation framework.

Final Guidance

The best tool stack is the one you'll use consistently. Start small, prove value, expand strategically. Expensive doesn't mean better. Sophistication doesn't equal results.

Most agents will likely get 80% of AI's value from:

- One premium LLM (ChatGPT or Claude)
- One property marketing tool (Virtual Staging AI or Reimaginehome)
- One design tool (Canva)
- One automation platform (Zapier or Make)

Total investment: $50-100/month

Everything beyond that is optimization, not transformation.

Choose tools that align with your business priorities, budget reality, and technical comfort level. Review this appendix quarterly as your needs evolve and new tools emerge.

Appendix C: Glossary of Terms

This glossary provides clear definitions of key artificial intelligence and real estate technology terms used throughout this book. Terms are organized by category for easy reference.

AI Fundamentals

Algorithm

A set of rules or instructions a computer follows to solve a problem or complete a task. In real estate AI, algorithms determine which properties to recommend to buyers, which leads to prioritize, or how to score the likelihood of a homeowner selling. The quality of the algorithm determines the quality of AI recommendations.

Artificial Intelligence (AI)

Technology that enables computers to perform tasks that typically require human intelligence, such as understanding language, recognizing patterns, making decisions, and learning from experience. In real estate, AI handles repetitive cognitive tasks (writing descriptions, analyzing data, generating content) so agents can focus on activities requiring human judgment and relationships.

Computer Vision

The AI capability that enables systems to interpret and understand visual information from images, videos, or live camera feeds. In real estate, computer vision can automatically analyze property photos to identify features (hardwood floors, granite countertops, pool), assess condition, flag quality issues, or even suggest optimal staging arrangements. Some platforms use it to verify listing photo accuracy or extract property details from images without manual data entry.

Context Window

The maximum amount of text an AI model can process and remember within a single conversation or task. Measured in tokens (roughly equivalent to words). Larger context windows allow AI to maintain coherence across longer documents or conversations. Important for real estate applications like analyzing lengthy contracts, processing multiple property comparables simultaneously, or maintaining context across extended client conversation threads. When exceeded, the AI "forgets" earlier parts of the conversation.

Generative AI

AI systems that create new content (text, images, designs, or other media) based on patterns learned from training data. ChatGPT generating property descriptions, DALL-E creating marketing graphics, or AI producing virtual staging are all examples of generative AI in action.

Hallucination (AI)

When an AI model generates false, fabricated, or nonsensical information while presenting it confidently as fact. Common in situations where the AI lacks relevant training data or misinterprets context. In real estate, this might manifest as inventing property features, fabricating market statistics, or creating fictional comparable sales. Always verify AI-generated factual claims, especially for client-facing content, pricing recommendations, or market analysis.

Large Language Model (LLM)

Advanced AI systems trained on vast amounts of text data to understand and generate human-like language. Examples include ChatGPT, Claude, and Gemini. Different LLMs have different strengths: some excel at creative writing, others at analytical tasks or following complex instructions. Most AI applications in this book leverage LLMs as the core technology.

Machine Learning

A subset of AI where systems learn and improve from data without being explicitly programmed for every scenario. For example, an AI system analyzing thousands of real estate transactions to identify patterns that predict which homeowners are most likely to sell, or which property features drive highest sale prices in specific neighborhoods.

Natural Language Processing (NLP)

Technology that enables computers to understand, interpret, and generate human language. This powers AI tools that write property descriptions, analyze client emails for sentiment, summarize lengthy contracts, or respond to inquiries in natural conversational language. The quality of NLP determines how "human" AI-generated content feels.

Prompt

The instruction or question you give to an AI system. The quality and specificity of your prompt directly determines the quality of AI output. Vague prompts produce generic results. Specific prompts with clear context, constraints, and examples produce professional-quality results. Prompt engineering (the skill of crafting effective prompts) is the most important AI skill for real estate professionals.

Prompt Engineering

The skill of crafting effective instructions for AI systems to get optimal results. Good prompt engineering includes providing context, specifying desired format and tone, including relevant details, and setting clear constraints. This is a learnable skill that dramatically improves AI output quality.

Temperature (AI Setting)

A parameter controlling randomness and creativity in AI outputs. Lower temperature (0.0-0.3) produces consistent, predictable, factual responses. Higher temperature (0.7-1.0) generates more creative, varied outputs. For real estate: use low temperature for contracts, CMAs, and factual content; use higher temperature for creative marketing copy and brainstorming sessions. Most consumer AI tools set temperature automatically, but understanding the concept helps diagnose why outputs vary.

Token

The basic unit AI models use to process text, roughly equivalent to 3-4 characters or about 0.75 words. AI systems have token limits (context windows) determining how much text they can process at once. Understanding tokens helps explain AI pricing models and why extremely long documents may need to be processed in sections.

Training Data

The information used to teach an AI system how to perform tasks. The quality, quantity, and recency of training data directly impact AI performance. Most LLMs have a knowledge cutoff date (the last date their training data includes), meaning they don't automatically know current events, recent market changes, or new regulations without you providing that information.

AI Tool Categories

Agentic AI

Autonomous AI systems that can execute complex, multi-step tasks and make decisions within defined parameters without constant human supervision. Unlike simple chatbots that respond to single prompts, agentic AI can pursue goals, handle multiple sequential steps, adapt to obstacles, and operate workflows end-to-end. Expected to become mainstream in real estate by 2028-2030 for functions like transaction coordination, lead nurturing, and marketing campaign management.

Chatbot

An AI-powered conversational interface that can interact with users via text or voice. In real estate, chatbots can qualify leads on your website 24/7, answer common questions, schedule showings, or provide property information when you're unavailable.

Computer Vision

AI technology that enables computers to understand and analyze images and video. In real estate, computer vision powers virtual staging, automatic photo enhancement, property feature detection from images, and quality assessment of listing photos.

Conversational AI

More sophisticated than basic chatbots, these systems understand context, remember previous interactions, and generate natural human-like responses. Modern conversational AI (like ChatGPT, Claude, or Gemini) can handle complex, multi-turn conversations and assist with detailed real estate tasks.

Multi-modal AI

AI systems that can process and generate multiple types of content-text, images, audio, video-within a single platform or interaction. For example, analyzing a property photo while generating a description, or processing a video walkthrough to create written highlights. Enables more sophisticated workflows like "analyze these listing photos and generate description, social posts, and email campaign based on what you see."

Voice AI

AI systems that process spoken language inputs and generate spoken responses, enabling hands-free interaction. In real estate, this includes using voice commands to update your CRM during showings, dictate client notes while driving, generate property descriptions through conversation, or respond to client texts via voice-to-text AI. Particularly valuable for agents who spend significant time in the field away from computers.

Common Challenges & Solutions

Change Management

The process of helping individuals, teams, or organizations adopt new technologies and workflows. Successful AI implementation requires addressing resistance, providing training and support, demonstrating value, and allowing time for adaptation.

Knowledge Cutoff

The date after which an AI model has no information, because training data only includes information up to that point. For example, a model with a December 2024 cutoff doesn't know events, data, or developments from 2025. Always verify time-sensitive information (interest rates, recent sales, new regulations, market shifts) from current authoritative sources rather than relying on AI knowledge.

Prompt Refinement

The iterative process of improving your AI instructions based on output quality. Initial prompts rarely produce perfect results. Review outputs, identify gaps or issues, adjust your prompt with more specificity or context, and test again. Save successful refined prompts for reuse.

Quality Control

Systematic review processes ensuring AI outputs meet your professional standards before client delivery. Include verification of facts, compliance checks, brand voice alignment, and accuracy confirmation. Quality control is your professional responsibility regardless of how content was created.

Data & Analytics Terms

A/B Testing

Comparing two versions of something (email subject lines, property descriptions, social media posts) to determine which performs better. AI tools can help generate variations for testing, and data from A/B tests helps refine your AI prompts for optimal results. Data-driven improvement beats guessing.

Conversion Rate

The percentage of leads that become clients, or any other desired action completion rate. Tracking conversion rates by lead source helps identify which marketing channels deliver the highest-quality prospects, allowing strategic resource allocation. AI should improve conversion rates through faster response times and more personalized follow-up.

KPI (Key Performance Indicator)

Measurable values that indicate how effectively you're achieving business objectives. Essential KPIs for AI implementation include time saved per week, lead conversion rate, client satisfaction scores, response time, and ROI on AI investments.

ROI (Return on Investment)

The financial return generated relative to the cost of an investment. For AI tools, calculate ROI by comparing value generated (time saved × your hourly value + additional revenue enabled) against total costs (subscriptions + implementation time). Most effective AI implementations deliver 500-1000%+ ROI within 6 months.

Sentiment Analysis

Using AI to determine the emotional tone of text (positive, negative, neutral). Applied to client communications, social media comments, or property reviews, sentiment analysis helps gauge satisfaction, identify concerns, and prioritize responses based on urgency and emotional state.

Single Source of Truth

A centralized, authoritative data repository that eliminates duplication and ensures all systems reference the same accurate information. In real estate practice, this typically means one CRM or database that all other tools and workflows connect to, rather than maintaining separate, potentially conflicting records across multiple platforms. Critical for AI success because algorithms trained on fragmented, duplicate, or outdated data produce unreliable results.

Ethics & Compliance

Algorithmic Bias

When AI systems produce prejudiced results due to biased training data or flawed algorithms. In real estate, this can manifest as discriminatory property recommendations, biased lead scoring, or Fair Housing violations in automated marketing. Regular audits of AI outputs for bias are essential professional responsibility.

Data Privacy

Protection of personal and confidential information from unauthorized access or misuse. When using AI tools, never input sensitive client data (financial information, Social Security numbers, confidential transaction details) into free public AI systems. Use paid, professional-grade tools with proper data protection agreements when working with client information.

Disclosure (AI context)

The requirement to inform clients about the use of AI in your services and clearly label AI-generated or AI-enhanced content. Virtual staging must be disclosed in all marketing materials. Significant photo enhancements should be disclosed. Professional responsibility standards require transparency about AI's role in your service delivery.

Fair Housing Act

Federal law prohibiting discrimination in housing based on race, color, national origin, religion, sex, familial status, or disability. AI tools must be carefully monitored to ensure they don't inadvertently violate Fair Housing through biased language, discriminatory targeting, or exclusionary marketing practices. Algorithmic bias in AI-generated content or targeting can create legal liability even when unintentional.

Implementation & Tools

API (Application Programming Interface)

A way for different software applications to communicate and share data with each other. APIs enable workflow automation by connecting your CRM, AI tools, email platform, and other systems so they work together seamlessly. When evaluating tools, ask whether they have API access—it's essential for building sophisticated automated workflows.

Automation Platform

Software designed to connect multiple applications and automate workflows between them. Examples include Make.com, Zapier, and n8n. These platforms allow you to build sophisticated multi-step automation without coding, orchestrating how your CRM, AI tools, email, and other systems work together.

Cloud-Based

Software and data stored on remote servers accessed via the internet, rather than installed on your local computer. Cloud-based AI tools offer advantages: access from any device, automatic updates, no local storage requirements, and collaboration capabilities. Most modern AI tools are cloud-based by design.

Integration

The connection between different software tools that allows them to share data and work together. Strong integrations eliminate manual data entry, reduce errors, and enable sophisticated automated workflows. When building your AI tech stack, prioritize tools that integrate well with your existing systems, particularly your CRM.

SaaS (Software as a Service)

Software delivered via the internet on a subscription basis rather than purchased and installed. Most AI tools operate on a SaaS model with monthly or annual pricing. Benefits include regular updates, lower upfront costs, and ability to cancel if tools don't deliver value.

Workflow Automation

Using software to automatically execute a series of tasks based on predefined rules and triggers, without requiring manual intervention at each step. For example, a new lead triggers automatic data enrichment, lead scoring, personalized email generation, CRM update, and agent notification—all happening in seconds. Workflow automation multiplies agent capacity by handling routine processes consistently.

Real Estate AI Applications

Answer Engine Optimization (AEO)

Strategy for ensuring your content, expertise, and business information is discoverable and cited when AI systems (like ChatGPT, Claude, or Perplexity) answer user queries. Unlike traditional SEO (optimizing for search engines), AEO focuses on being the authoritative source AI models reference when users ask questions like "who is the best investor-friendly agent in Austin?"

CRM (Customer Relationship Management)

Software for managing interactions with clients and prospects, tracking communications, deals, and relationship history. Your CRM should be your single source of truth—the authoritative database that all other tools and workflows connect to. Many modern CRMs include AI features for task prioritization, automated follow-up, and lead intelligence.

Data Enrichment

The process of enhancing existing contact or property data by adding information from external sources. For real estate agents, this might mean taking a basic lead (name, email) and automatically adding property ownership history, home value estimates, recent permits filed, LinkedIn profile, employment information, and local connections. Enrichment transforms sparse contact lists into detailed dossiers that enable highly personalized, relevant outreach.

Drip Campaign

A series of automated emails or messages sent to leads or clients over time, designed to nurture relationships, provide value, and stay top-of-mind. AI enhances drip campaigns by personalizing content based on recipient behavior and preferences, dramatically improving engagement and conversion rates compared to generic sequences.

GTM Engineering (Go-To-Market Engineering)

A systematic approach to business development that combines data sourcing, enrichment platforms, and AI-powered personalization to identify prospects showing intent signals and deliver timely, contextually relevant outreach. In real estate, this might involve monitoring building permits, property ownership changes, or business expansions to identify potential sellers or buyers at the exact moment they're most receptive to contact. Requires orchestrating multiple specialized tools (data sources, enrichment platforms, AI generation, automation) into cohesive workflows.

Lead Scoring

Using AI to automatically evaluate and rank leads based on their likelihood to convert into clients. Factors might include behavior (properties viewed, emails opened), demographics, timing indicators, and engagement patterns. Effective lead scoring helps agents prioritize their time on the highest-potential prospects, dramatically improving conversion rates.

Predictive Analytics

Using AI and machine learning to analyze historical data and identify patterns that predict future outcomes. In real estate, this includes predicting which homeowners are likely to sell, which neighborhoods will appreciate fastest, or which leads are most likely to convert. Accuracy improves with data quality and quantity.

Sentiment Analysis

An AI technique that evaluates text to determine the emotional tone, attitude, or opinion expressed. For real estate agents, sentiment analysis can scan client emails, text messages, or social media comments to gauge satisfaction levels, detect urgency or frustration, and prioritize responses accordingly. It helps you understand not just what clients are saying, but how they feel about properties, your service, or the buying/selling process-allowing you to respond with appropriate empathy and timing.

Virtual Staging

Using AI to digitally furnish and decorate empty rooms in property photos, helping buyers visualize the space's potential. Much more affordable than physical staging ($15-40 per listing vs. $2,000-5,000) but requires clear disclosure in all marketing materials per NAR guidelines.

Waterfall Enrichment

A sequential data lookup strategy where the system checks multiple data sources in priority order until the needed information is found. If Source A doesn't have a prospect's email address, automatically check Source B, then Source C, etc. Maximizes data coverage while minimizing costs (expensive sources only queried when cheaper ones fail).

Workflow Productivity

Batch Processing

Completing similar tasks together in one focused session rather than scattering them throughout the week. Use AI to batch-create content (generating a week's social posts at once, writing multiple property descriptions in one sitting) for greater efficiency and consistency.

Scalability

The ability to handle increased workload without proportional increases in time or resources. AI-powered workflows are highly scalable-serving 30 clients with AI assistance takes similar effort to serving 20, whereas manual processes require proportionally more time for each additional client.

Standard Operating Procedure (SOP)

Documented step-by-step instructions for completing a task consistently. Creating SOPs for your AI-assisted workflows ensures quality, enables training others, and provides reference when troubleshooting issues.

Time Blocking

Scheduling dedicated time periods for specific activities to improve focus and productivity. AI tools work best when you block time for content creation, automation setup, or review sessions rather than trying to use them in fragmented moments between other tasks. Protect your peak energy time for high-value work; use AI to handle routine tasks during lower-energy periods.

---

www.ingramcontent.com/pod-product-compliance
Lightning Source LLC
Chambersburg PA
CBHW071542210326
41597CB00019B/3091